Dr. Fanatomy's

HUMAN ANATOMY AND NEUROANATOMY COLORING BOOK

WITH FACTS & MCQS

(MULTIPLE CHOICE QUESTIONS)

2-IN-1 BOOK

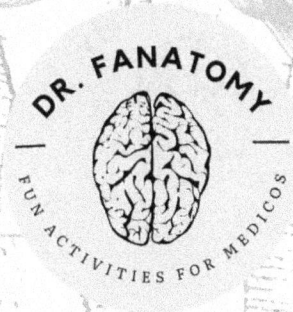

DR. FANATOMY

FUN ACTIVITIES FOR MEDICOS

Book -1

HUMAN ANATOMY AND PHYSIOLOGY COLORING BOOK

WITH FACTS & MCQS

(MULTIPLE CHOICE QUESTIONS)

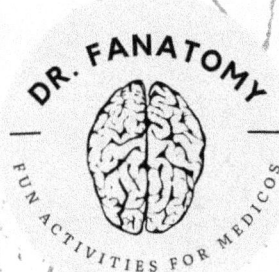

DR. FANATOMY

FUN ACTIVITIES FOR MEDICOS

Chapters

"Anatomy is the great ocean of intelligence upon which the true physician must sail. Bacteriology is but one little harbor."

— John E. Link

- This book is a part of the Dr. Fanatomy Series.
- This colored book has 127 pages and 45 coloring pages.
- The colored photo of the illustration is also provided for inspiration.
- Facts related to the topic are provided for each coloring page.
- 30 Multiple-choice questions are provided at the end of the book.
- This is a unique book with all interiors in color and colored photos provided for each illustration.
- It is suggested to use colored pencils or light crayons.

Other Book you can check out

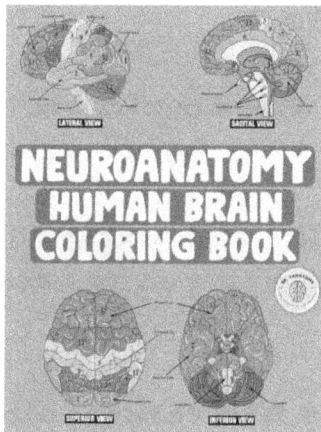

NEUROANATOMY HUMAN BRAIN COLORING BOOK

LATERAL VIEW · SAGITAL VIEW · SUPERIOR VIEW · INFERIOR VIEW

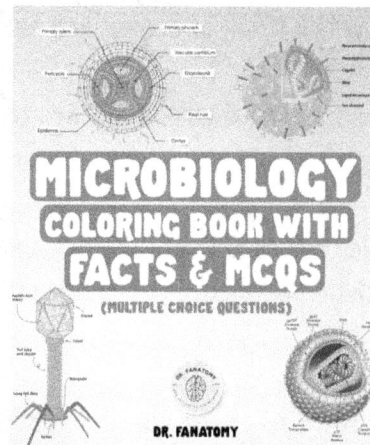

MICROBIOLOGY COLORING BOOK WITH FACTS & MCQS (MULTIPLE CHOICE QUESTIONS)

DR. FANATOMY

1. Introduction & History

- The human body is a combination of 11 interconnected systems.

- Each system performs one primary function or task.

- Systems are composed of primary components called organs, which are made up of tissues and cells.

ANATOMY & PHYSIOLOGY EXPLAINED:

- Anatomy studies body structure and its parts, including size, form, structure, and coloration.

- Photos of organs in anatomy are seen as exploded views and cross-sections, which gives us a clear understanding.

- Body organs move continually as we move, take a breath, sleep, and eat.

- Physiology is the study of organs and answers - "how do they function?"

- Both disciplines are interrelated since the practical function of an organ depends on exactly how it is made.

- For example, red blood cells include iron in particles of the protein called hemoglobin; this is an aspect of their anatomy. The presence of iron allows red blood cells to carry oxygen, which is their function.

- All cells in the body must obtain oxygen to operate appropriately, so the physiology of red blood cells is necessary to the physiology of the body.

FATHER OF MODERN ANATOMY

Andreas Vesalius

- Andreas Vesalius was a Belgian anatomist and physician, born in 1514 right into a household of physicians.

- He is known as the father of modern anatomy, and his work was the beginning of contemporary anatomy.

Major branches of Anatomy

Anatomy

Gross anatomy	Microscopic anatomy/ histology	Cell biology (Cytology) & cytogenetics.
Surface anatomy	Radiological anatomy.	Developmental anatomy/embryology.

Coloring Page – Andreas Vesalius

ORGAN SYSTEMS

SYSTEM	FUNCTION	ORGANS
Integumentary	Prevents water loss and saves us from pathogens and chemicals.	skin
Skeletal	Sustains the body and safeguards interior body organs. Gives structure to the muscles.	bones, ligaments
Muscular	Produces heat and moves skeleton.	muscles, tendons
Nervous	Analyzes sensory information and manages body functions.	brain, nerves, eyes, ears
Endocrine	Controls body functions and also manages daily metabolic processes using hormonal agents.	thyroid gland, pituitary gland, pancreas
Circulatory	Transports oxygen as well as nutrients to cells as well as eliminates waste items.	heart, blood, arteries

ORGAN SYSTEMS

SYSTEM	FUNCTION	ORGANS
Lymphatic	Provides immunity & destroys pathogens	lymph nodes
Respiratory	Exchanges oxygen and Co2 between the air & blood.	lungs, trachea, larynx,diaphragm
Digestive	Breaks down food to be absorbed by the body	stomach, colon, liver,pancreas
Urinary	Eliminates waste products from the blood and controls the PH of blood.	kidneys, urinary bladde
Reproductive	Produces eggs or sperm and place for embryo-fetus	prostate gland, ovaries, uterus

2.Levels of organization of Body

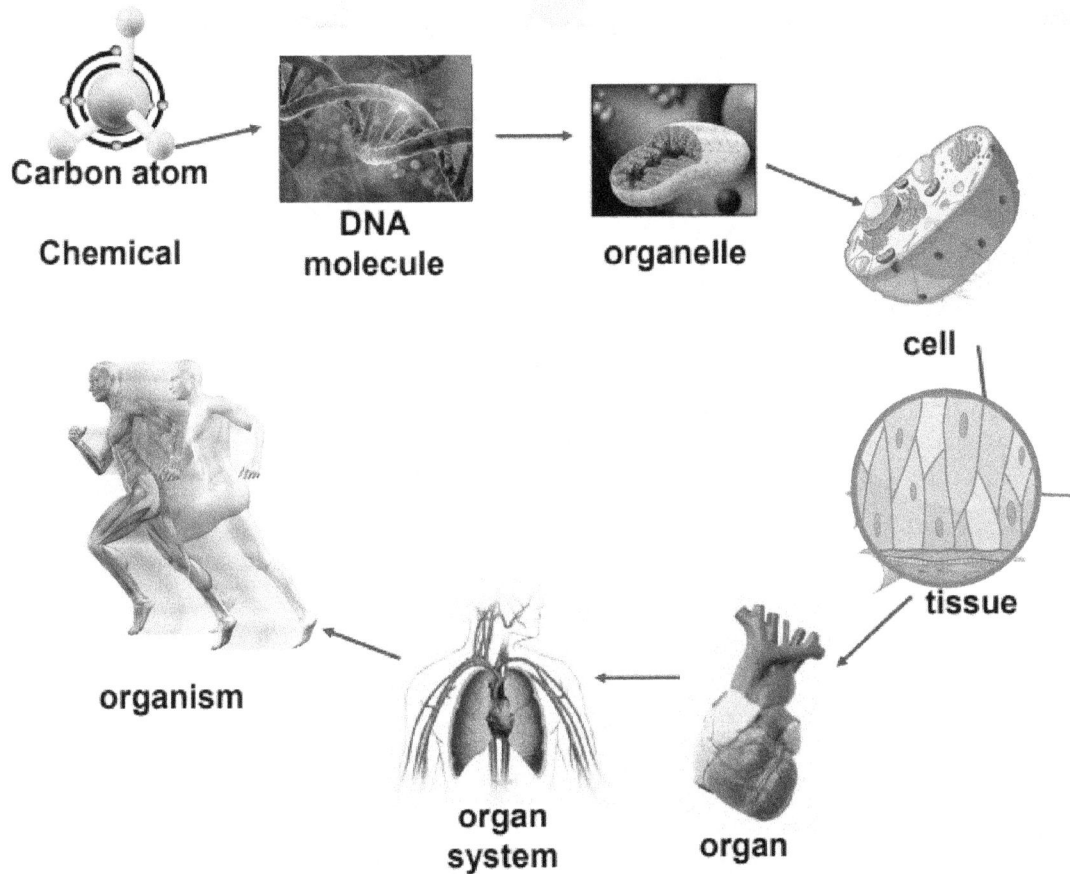

Carbon atom

Chemical

DNA molecule

organelle

cell

tissue

organism

organ system

organ

- The human body is categorized into structural and functional levels based on complexity. Each higher level incorporates the structures and functions of the previous level.

- You start with the most superficial level, the chemical level, and proceed to cells, tissues, organs, and organ systems.

Organization of a Human Body

Select different colors for each of the areas provided with a coding numbers and corresponding structures in the diagram.

1. Chemical
2. Cell
3. Tissue
4. Cell
5. Organ
6. Organ Systems
7. Organism

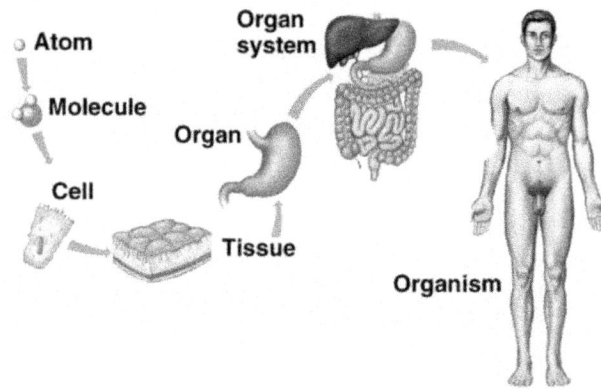

_____

_____

_____

_____

2 _____

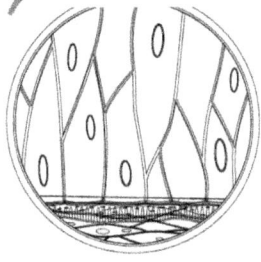

3 _____

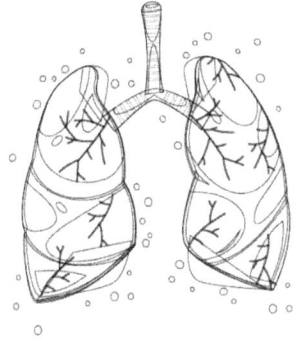

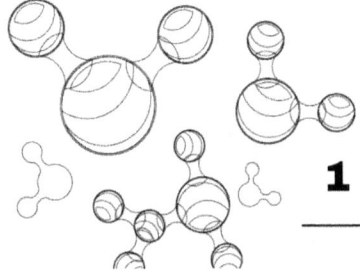

4 _____

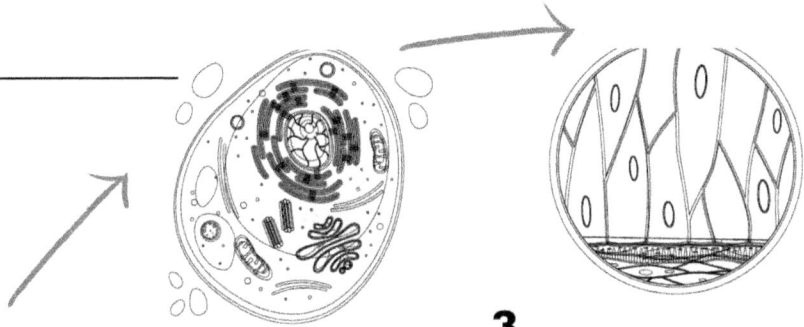

1 _____

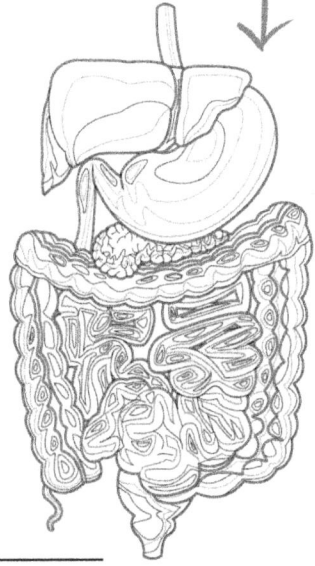

6 _____

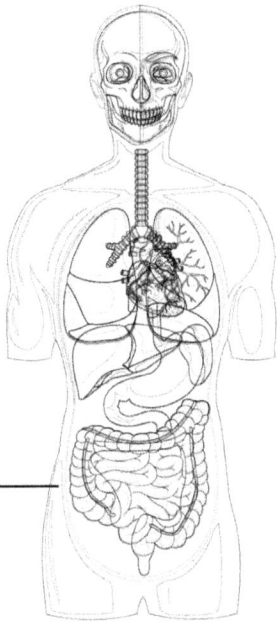

5 _____

Type of Cells

Select different colors for each of the areas provided with a coding numbers and corresponding structures in the diagram.

1. Blastocyst
2. Neuron
3. Enterocytes
4. Chondrocyte
5. Cardiac Cells
6. Epithelial Cells
7. Red Blood Cells
8. Fat Cells

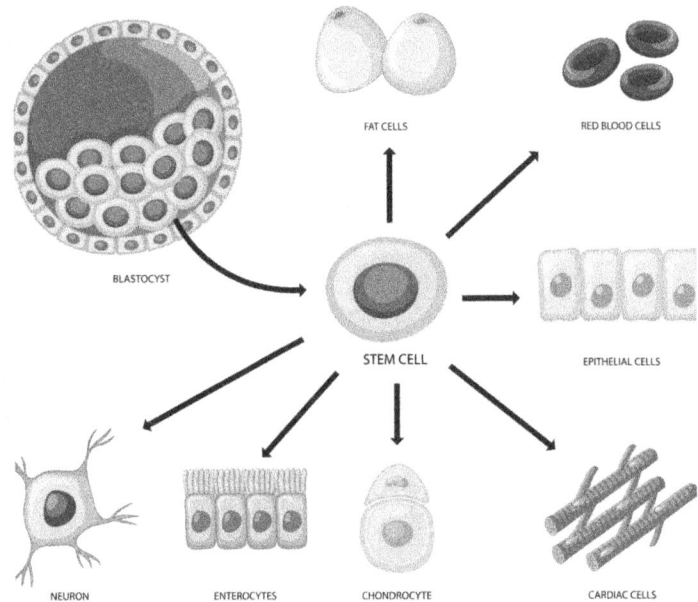

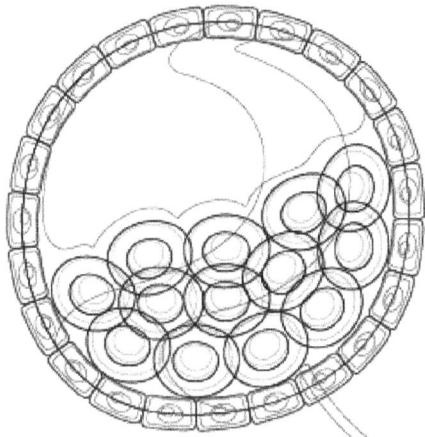

1. _____

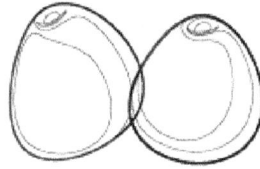

8. _____

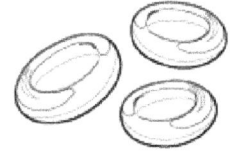

7. _____

STEM CELL

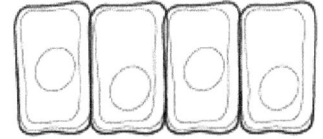

6. _____

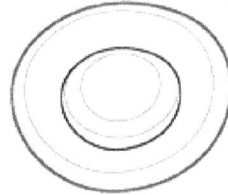

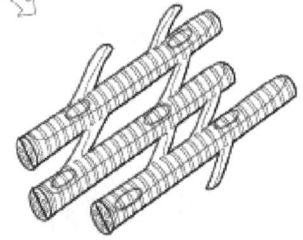

5. _____

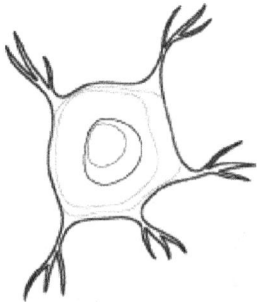

2. _____

3. _____

4.. _____

3. Body Cavities

Body Cavities

- A body cavity is an area of the human body that contains organs.
- It is lined with a layer of fluid-filled cells to shield the body organs from damage as they walk around.
- Body cavities create during development as solid masses of tissue fold inward on themselves, developing pockets.
- For example, the cranial cavity houses the human brain.

Cavity Types:

The body has two primary cavities:

- Dorsal cavity (posterior)
- Ventral cavity (anterior).

The dorsal cavity has the main nerves and includes the cranial cavity and the vertebral or spinal cavity.

The ventral cavity contains two compartments, the thoracic cavity and also the abdominal cavity, separated by the diaphragm.

Body Cavities

Select different colors for each of the areas provided with a coding numbers and corresponding structures in the diagram.

1. Cranial Cavity
2. Thoracic Cavity
3. Ventral Cavity
4. Abdominopelvic Cavity
5. Pelvic Cavity
6. Spinal Cavity
7. Dorsal Cavity

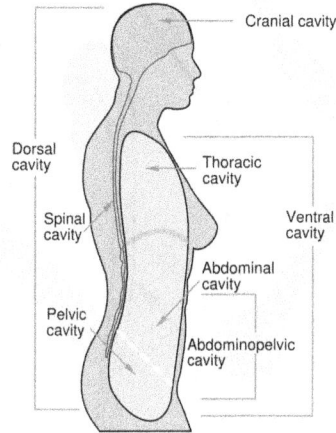

_____

_____

_____

_____

_____

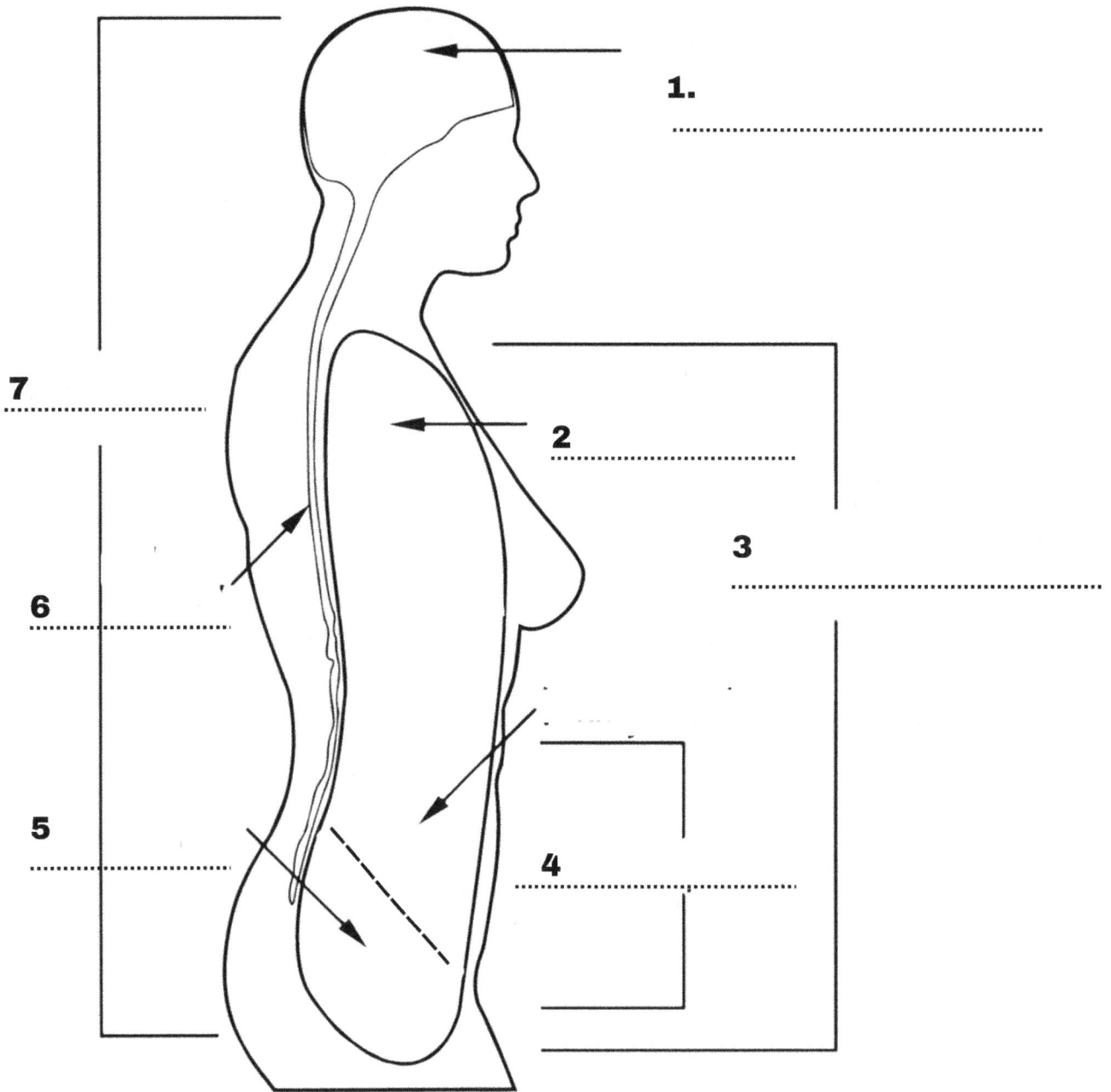

1.

....

2 ....

3

....

7 ....

6 ....

5 ....

4

....

4. The Integumentary System

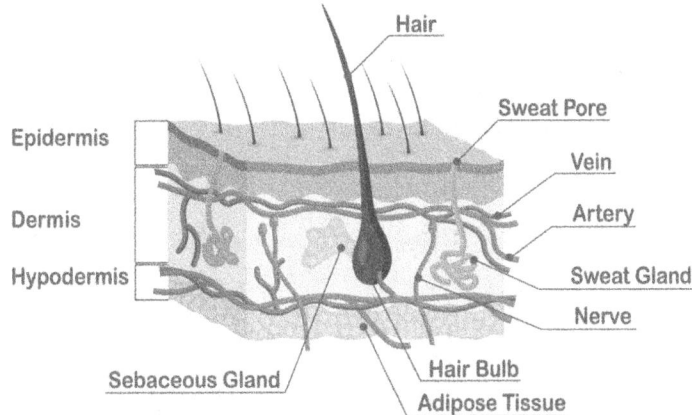

Hair

Epidermis

Dermis

Hypodermis

Sweat Pore

Vein

Artery

Sweat Gland

Nerve

Sebaceous Gland

Hair Bulb

Adipose Tissue

- The integumentary system contains the skin, its accessory structures such as hair and gland, and the subcutaneous tissue listed below the skin.

- It kinds the outside body covering and also protects deeper tissues from injury.

- It manufactures vitamin D and saves the body from pain, pressure, and so on.

- Major Organs:
  - Skin
  - Hair
  - Sweat Glands
  - Nails

The Integumentary System

Select different colors for each of the areas provided with a coding numbers and corresponding structures in the diagram.

1. Hair	5. Sweat Gland	9. Sebaceous Gland
2. Sweat Pore	6. Nerve	10. Hypodermis
3. Vein	7. Hair Bulb	11. Dermis
4. Artery	8. Adipose Tissue	12. Epidermis

_____

_____

_____

_____

_____

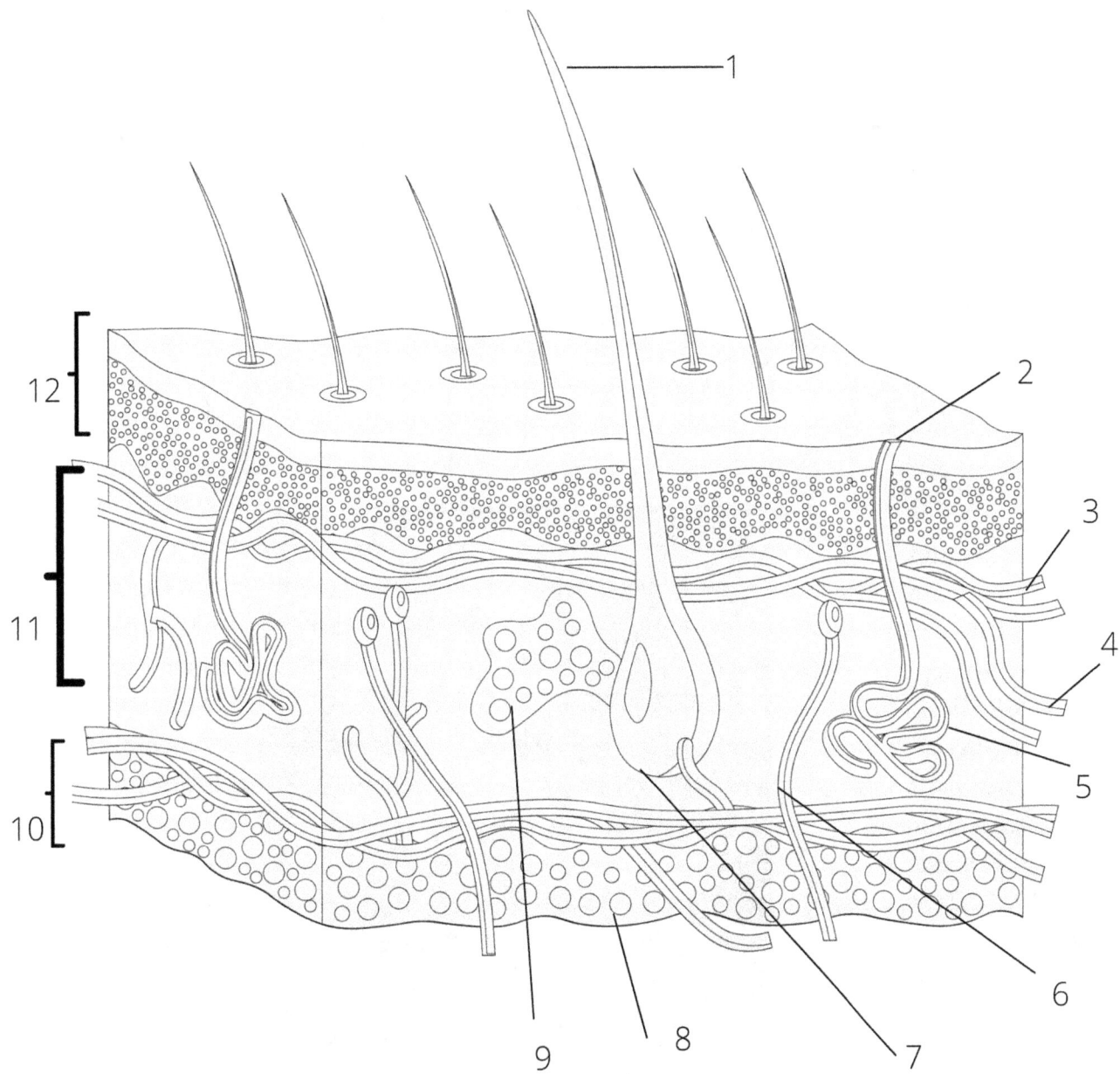

1

2

3

4

5

6

7

8

9

10

11

12

5. The Skeletal System

- The skeletal system is made up of bones and tendons.
- It offers a structure that sustains the body and muscles connected to bones.
- It safeguards inner organs from injury.
- It contains and also protects the red bone marrow, the main blood-forming tissue.

1. Cranium
2. Lower Maxilar
3. Ribs
4. Humerus
5. Radius
6. Ulna
7. Carpal Bones
8. Metacarpals
9. Phalages
10. Patela
11. Febula
12. Metatarsals

13. Clavicle
14. Scapula
15. Sternum
16. Vertebrae
17. Pelvis
18. Sacrum
19. Coccyx
20. Pelvis
21. Femur
22. Tibia
23. Tarsal Bones
24. Phalanges

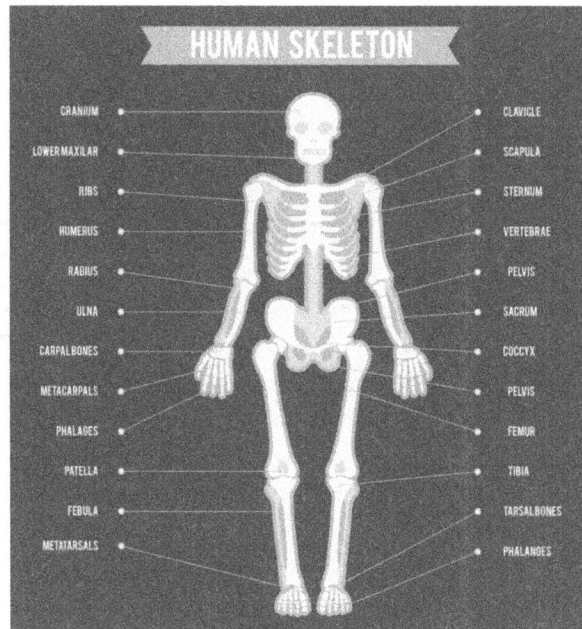

HUMAN SKELETON

CRANIUM
LOWER MAXILAR
RIBS
HUMERUS
RADIUS
ULNA
CARPAL BONES
METACARPALS
PHALAGES
PATELLA
FEBULA
METATARSALS

CLAVICLE
SCAPULA
STERNUM
VERTEBRAE
PELVIS
SACRUM
COCCYX
PELVIS
FEMUR
TIBIA
TARSAL BONES
PHALANGES

Skeletal System

HUMAN SKELETON

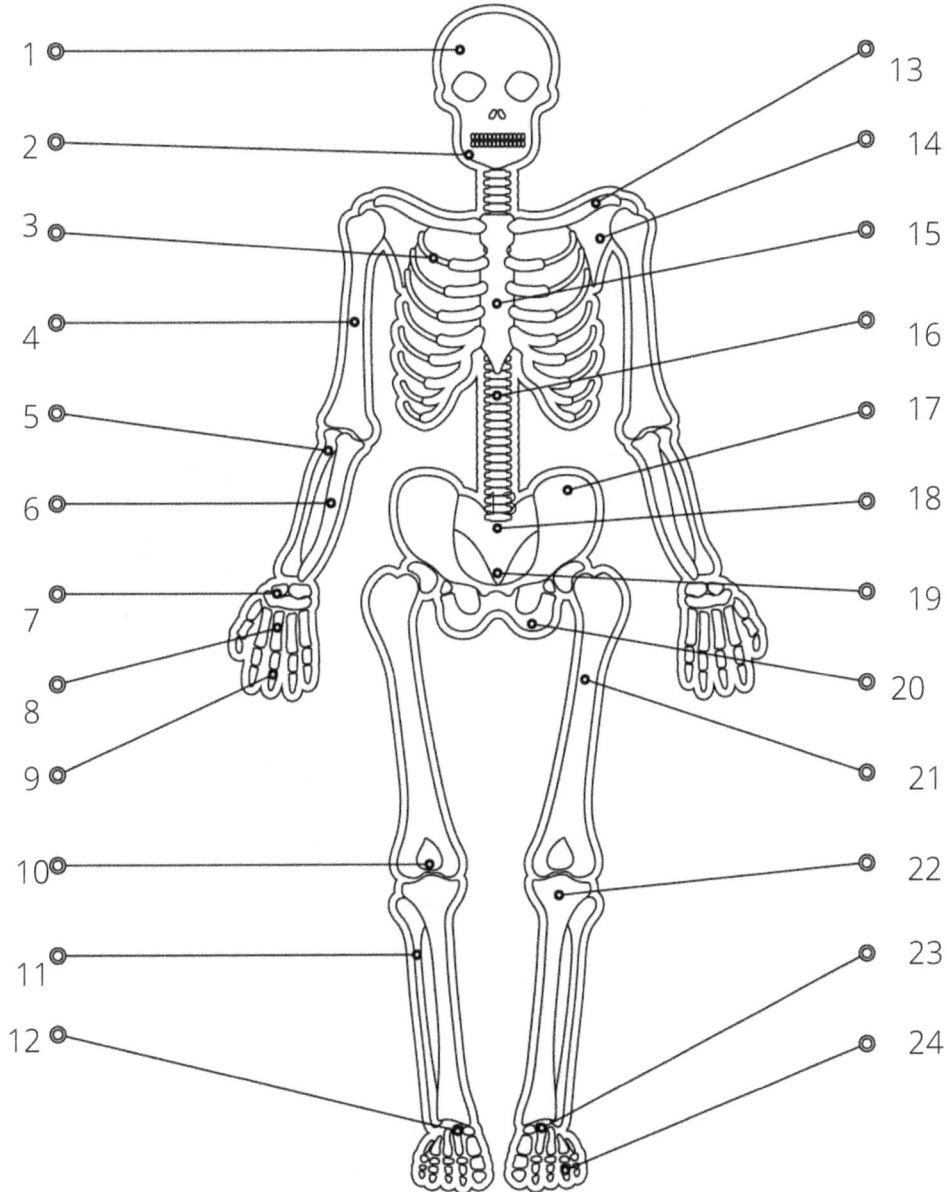

1
2
3
4
5
6
7
8
9
10
11
12

13
14
15
16
17
18
19
20
21
22
23
24

Skull

Select different colors for each of the areas provided with a coding numbers and corresponding structures in the diagram.

1. Frontal
2. Temporal
3. Zygomatic
4. Ethmoid
5. Mandible

6. Maxilla
7. Sphenoid
8. Parietal
9. Lacrimal
10. Nasal

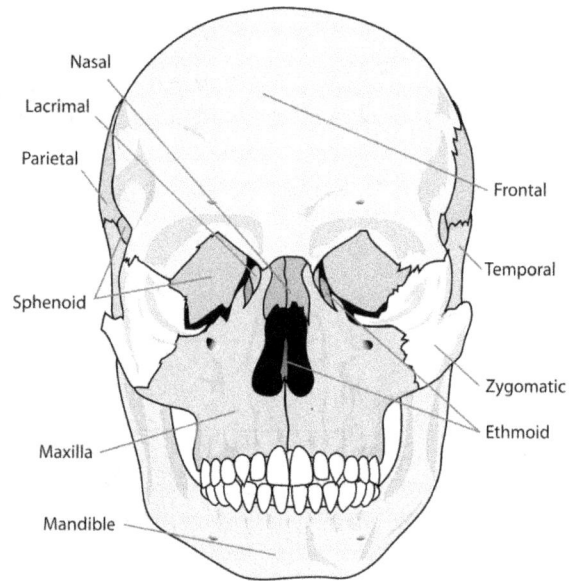

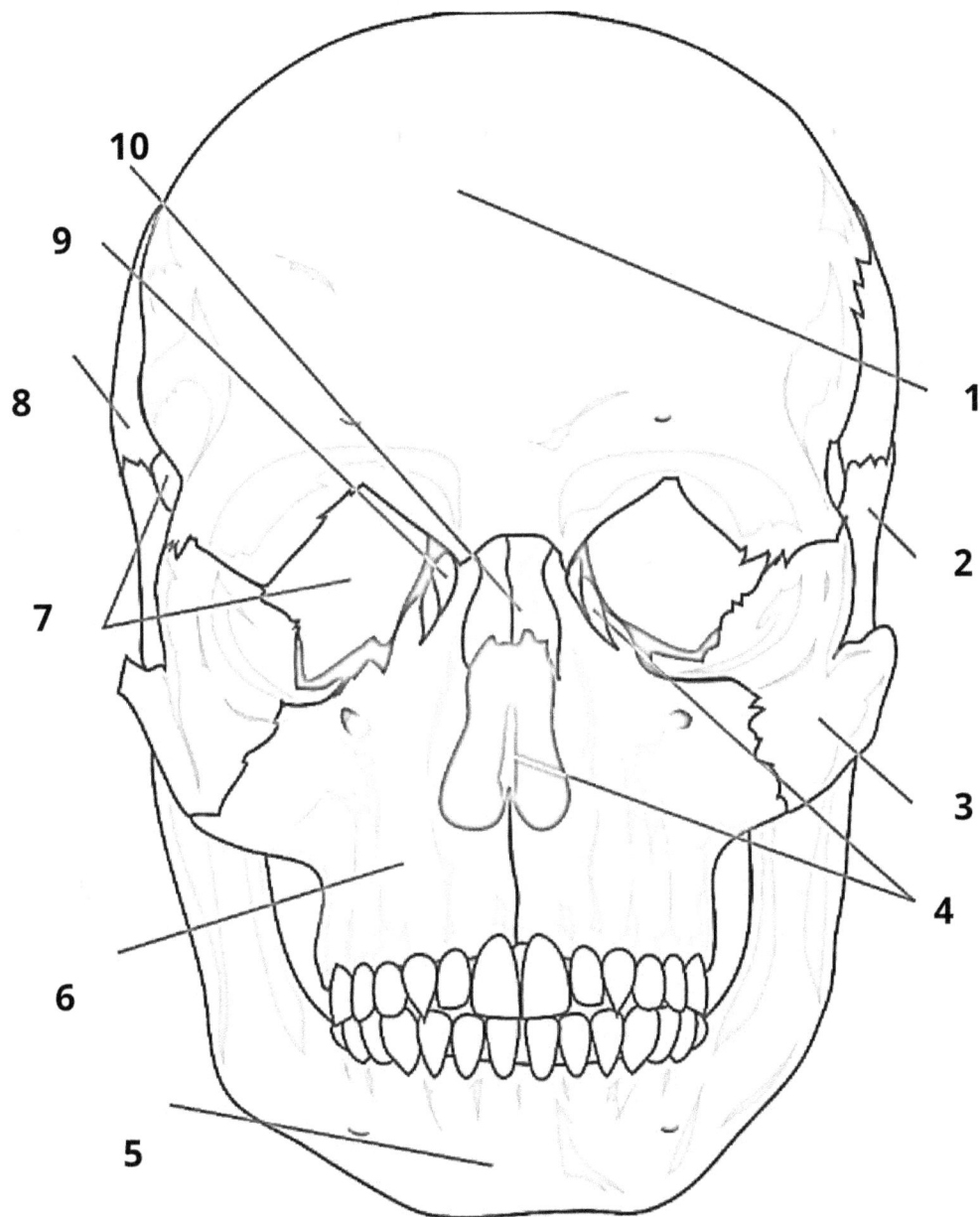

Hand Bone

Select different colors for each of the areas provided with a coding numbers and corresponding structures in the diagram.

1. Distal Phalanges
2. Middle Phalanges
3. Proximal Phalanges
4. Metacarpal Bones
5. Trapezoid

6. Trapezium
7. Scaphoid
8. Capitate
9. Radius
10. Ulna

11. Lunate
12. Triquetrum
13. Pisiform
14. Hamate
15. Hamate Hook

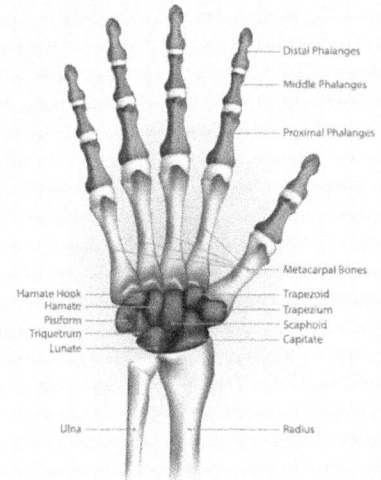

_____

_____

_____

_____

_____

_____

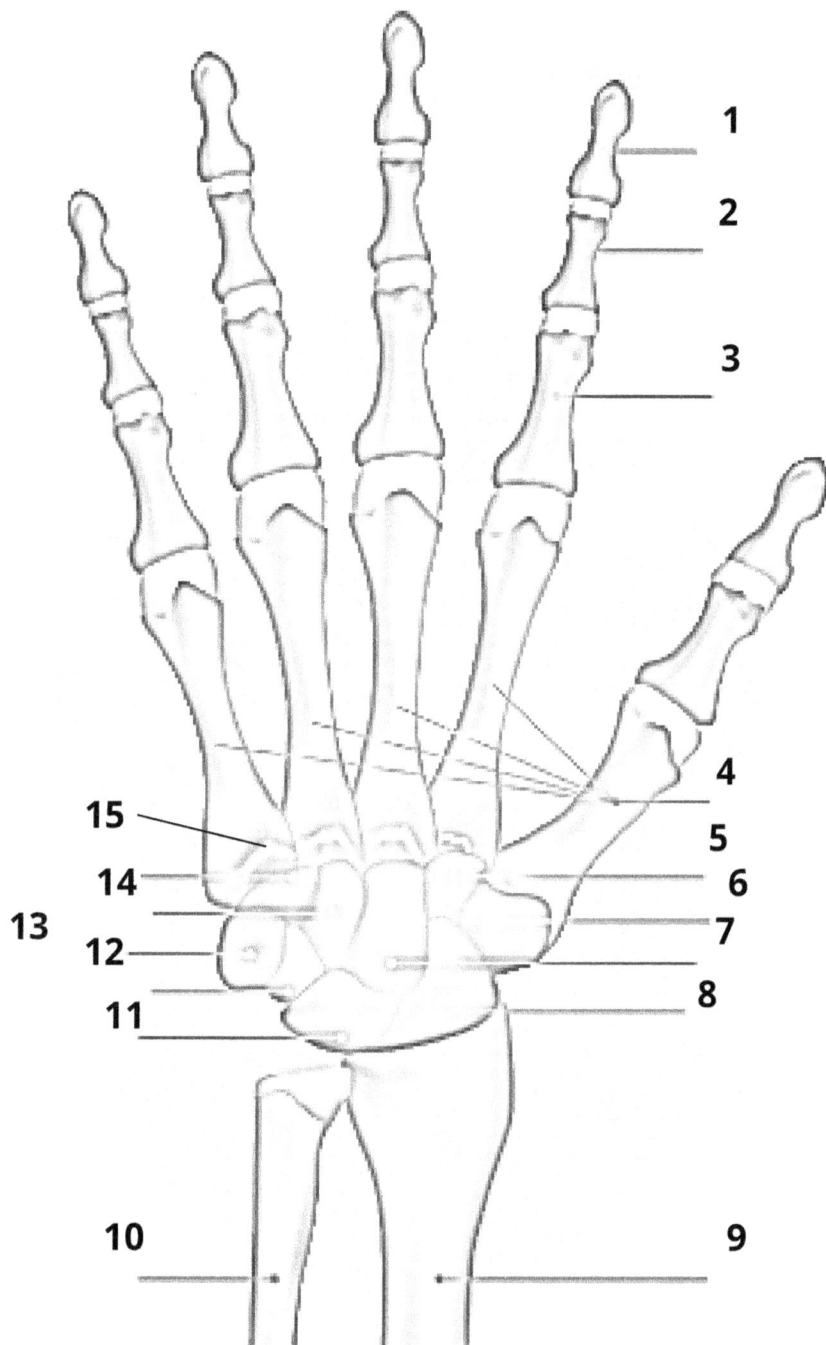

Vertebrae

Select different colors for each of the areas provided with a coding numbers and corresponding structures in the diagram.

Human Vertebrae Anatomy

1. Spinal Cord
2. Spinous Process
3. Nerve Root
4. Pedicle
5. Disc
6. Vertebral Body

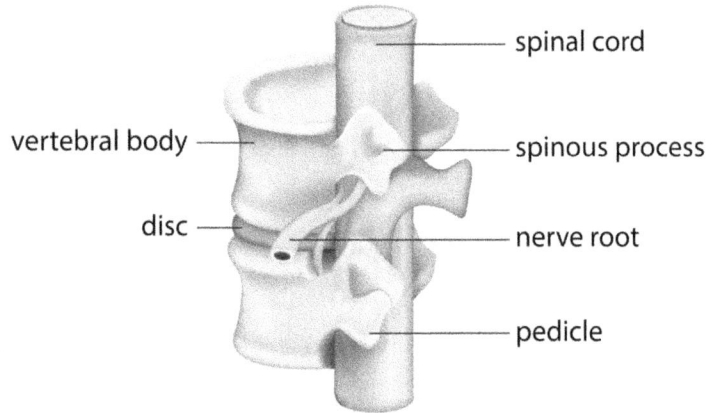

spinal cord

vertebral body

spinous process

disc

nerve root

pedicle

_____

_____

_____

_____

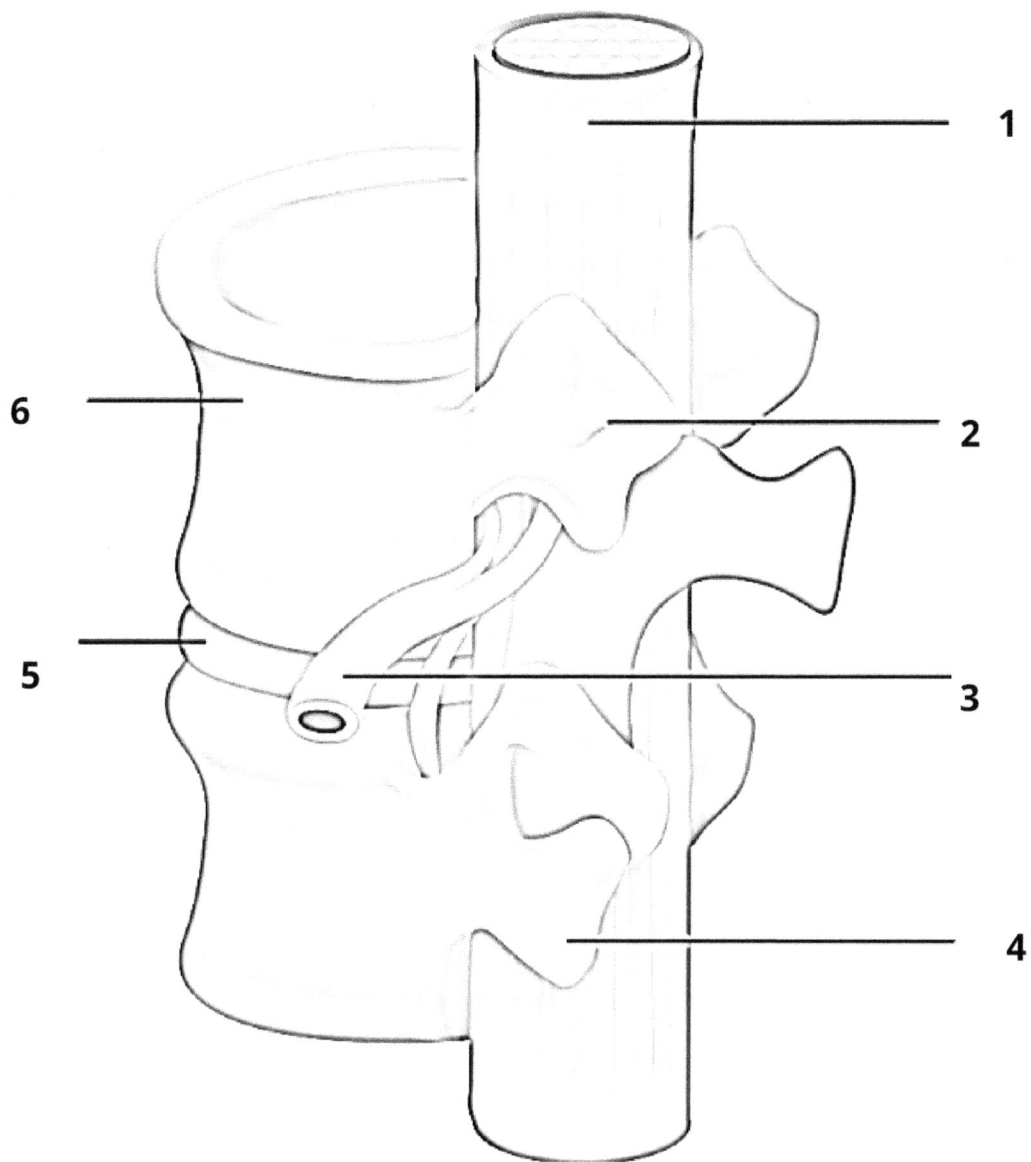

1

2

6

5

3

4

Foot Bone

Select different colors for each of the areas provided with a coding numbers and corresponding structures in the diagram.

1. Distal
2. Middle
3. Proximal
4. Metatarsals
5. Lateral Cuneiform
6. Cuboid
7. Talus
8. Navicular
9. Intermediate Cuneiform
10. Medial Cuneiform

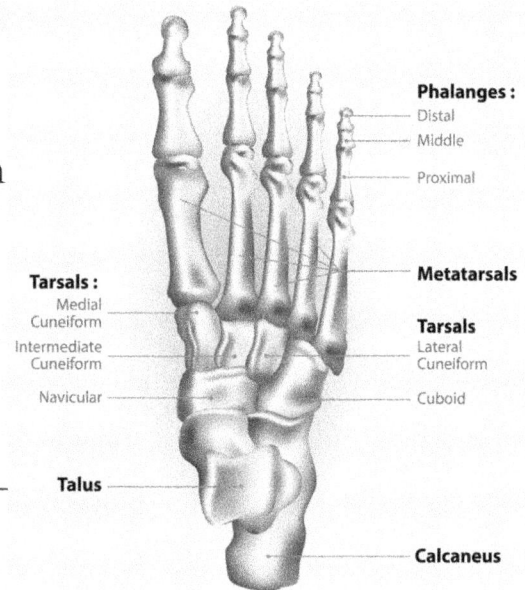

Phalanges :
Distal
Middle
Proximal

Metatarsals

Tarsals :
Medial Cuneiform
Intermediate Cuneiform
Navicular

Tarsals
Lateral Cuneiform
Cuboid

Talus

Calcaneus

_____

_____

_____

_____

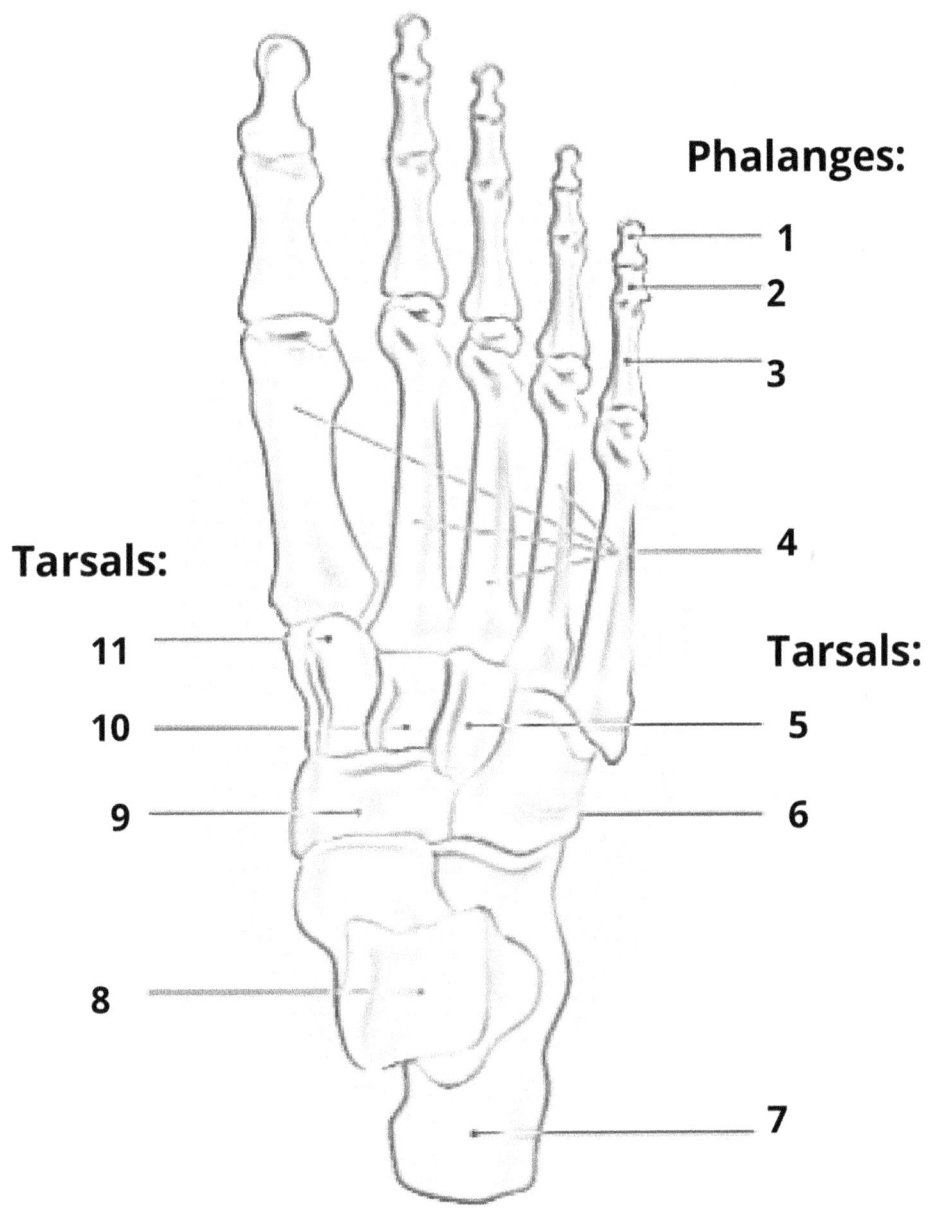

Phalanges:

1

2

3

Tarsals:

4

11

Tarsals:

10

5

9

6

8

7

Knee Bone

Select different colors for each of the areas provided with a coding numbers and corresponding structures in the diagram.

1. Femur
2. Lateral Meniscus
3. Lateral Collateral Ligament
4. Tibia
5. Medial Meniscus
6. Medial Collateral Ligament
7. Posterior crucuate ligament

Human Knee Anatomy

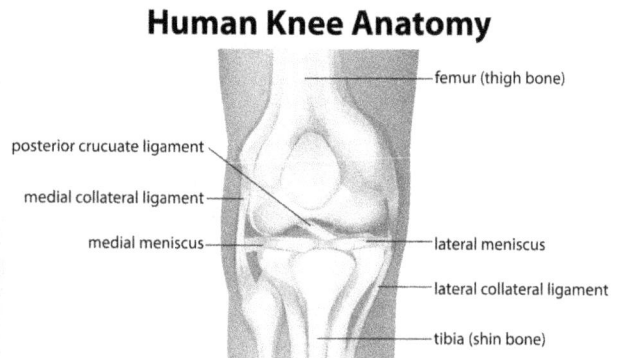

femur (thigh bone)

posterior crucuate ligament

medial collateral ligament

medial meniscus

lateral meniscus

lateral collateral ligament

tibia (shin bone)

_____

_____

_____

_____

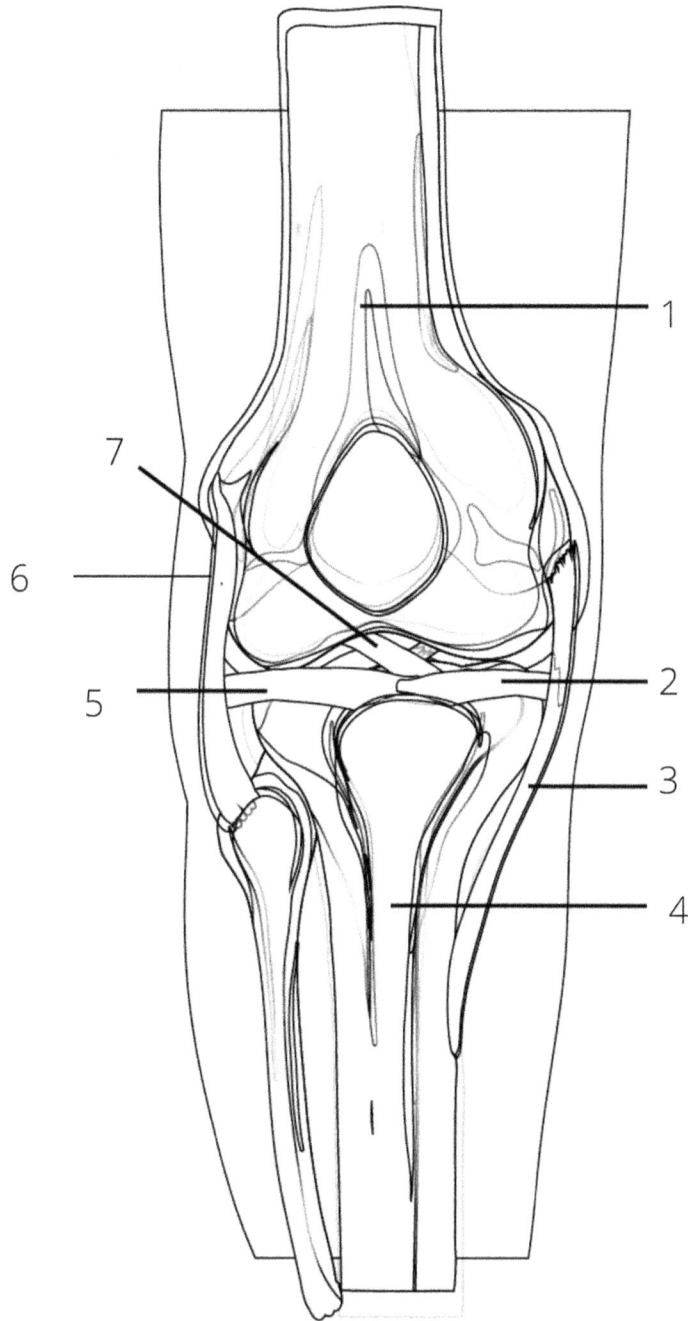

Pelvic Girdle(Hip Joint)

Select different colors for each of the areas provided with a coding numbers and corresponding structures in the diagram.

1. Iliac Crest
2. Tubercle Iliac Crest
3. Anterior Superior Iliac Spine
4. Anterior Inferior Iliac Spine
5. Iliopubic Eminence
6. Pectineal Line
7. Articular Cartilage
8. Greater Trochanter of Femur
9. Ischial Tuberosity
10. Inferior Pubic Ramus
11. Pubic Arch
   Obturator Foramen
12. Superior Pubic Ramus
13. Greater Sciatic Notch
14. Sacral Promontory
15. Sacrum

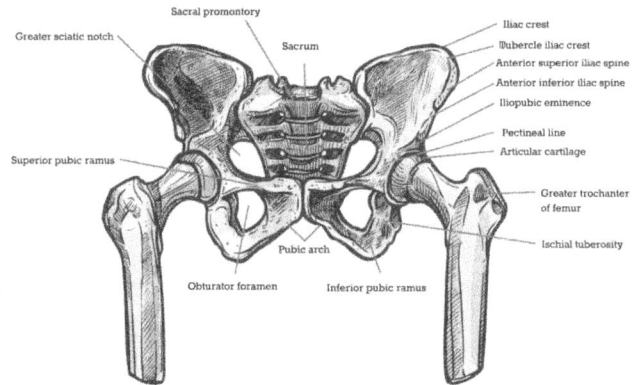

_____

_____

_____

_____

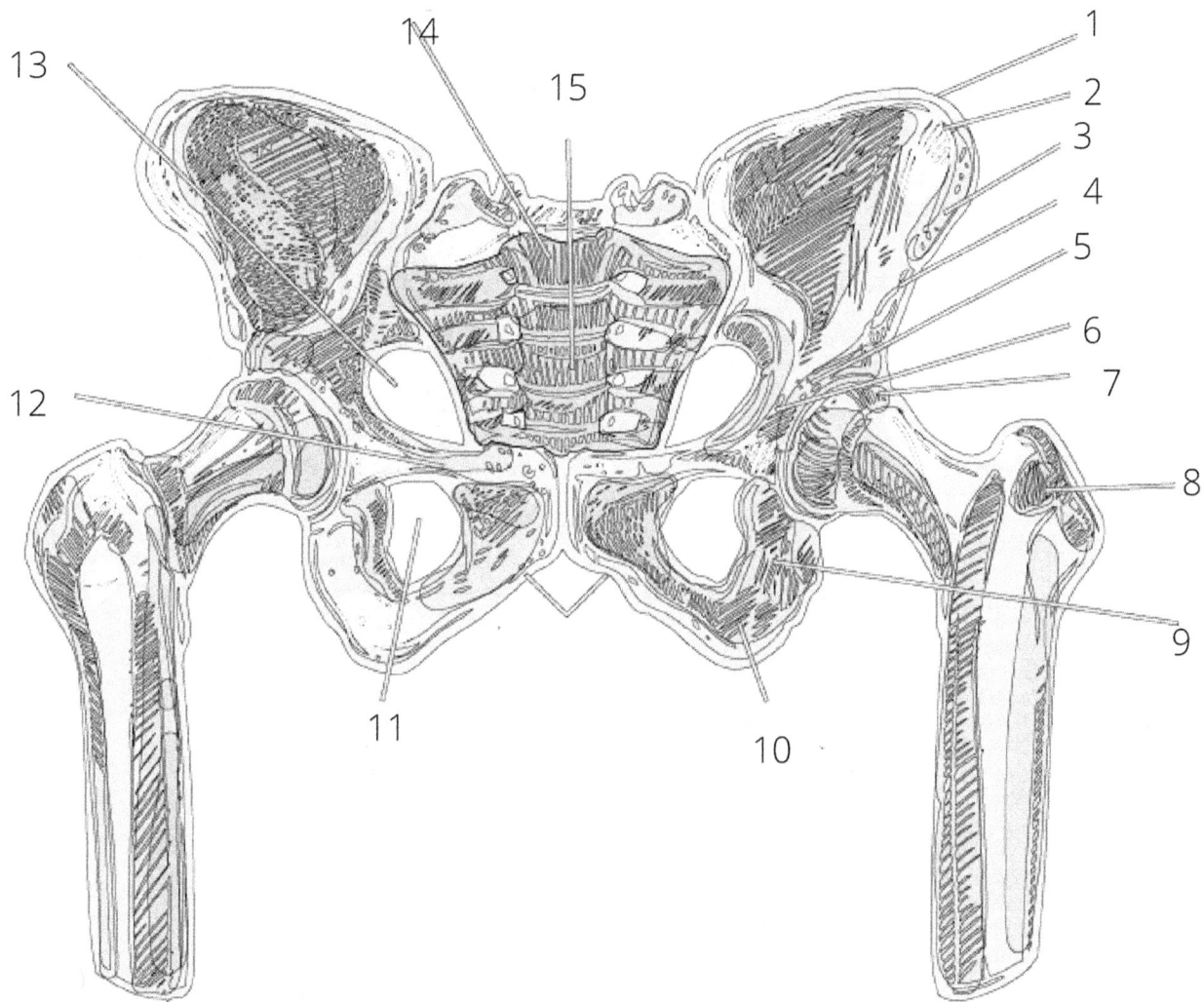

6. Muscular System

- The primary function of the muscular system is to move the skeleton. However, the muscle contractions required for the movement also produce heat, maintaining a constant body temperature.

- The muscular system is an organ system containing skeletal, smooth, and heart muscles. It allows body movement, keeps position, and distributes blood throughout the body.

- Attached to the bones of the skeletal system have to do with 700 named muscle masses that compose approximately fifty percent of a person's body weight.

Human Muscle Anatomy

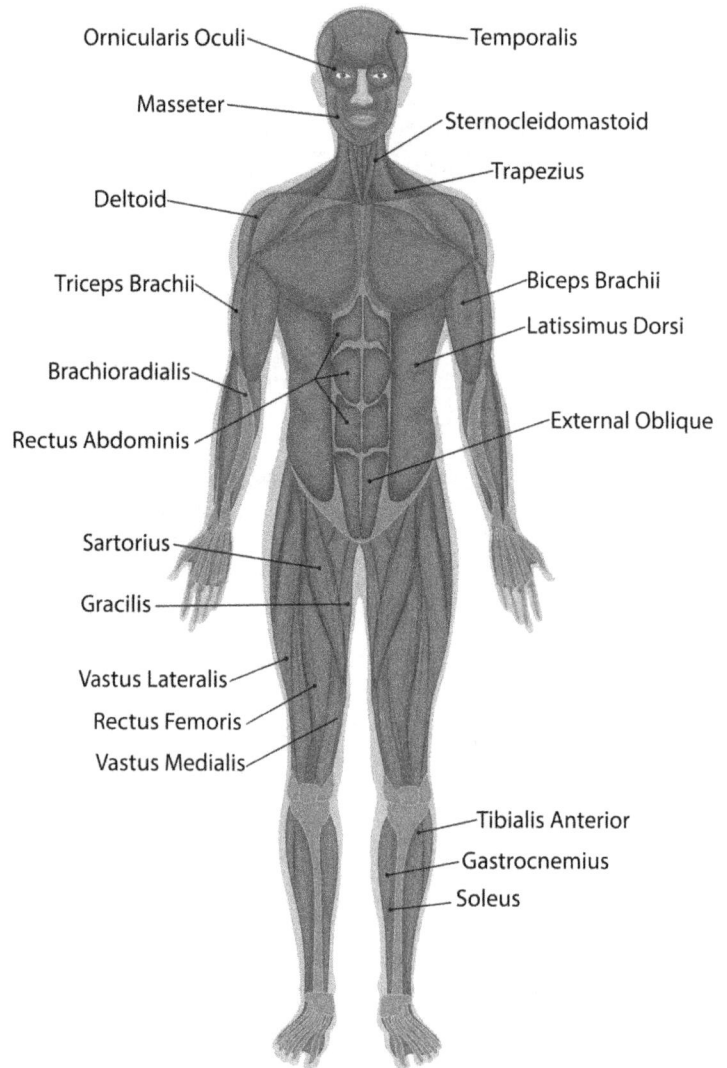

Ornicularis Oculi

Temporalis

Masseter

Sternocleidomastoid

Trapezius

Deltoid

Triceps Brachii

Biceps Brachii

Latissimus Dorsi

Brachioradialis

Rectus Abdominis

External Oblique

Sartorius

Gracilis

Vastus Lateralis

Rectus Femoris

Vastus Medialis

Tibialis Anterior

Gastrocnemius

Soleus

Human Muscle Anatomy

Select different colors for each of the areas provided with a coding numbers and corresponding structures in the diagram.

1. Skeletal Muscle
2. Epimysium
3. Fasciculus
4. Muscle Fascicles
5. Sarcolemma
6. Sarcoplasm
7. Actin Thin Filament
8. Myosin
9. Perimysium

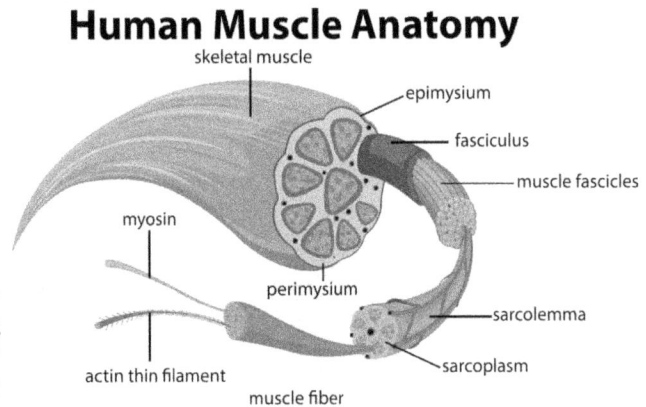

Human Muscle Anatomy

skeletal muscle

epimysium

fasciculus

muscle fascicles

myosin

perimysium

sarcolemma

sarcoplasm

actin thin filament

muscle fiber

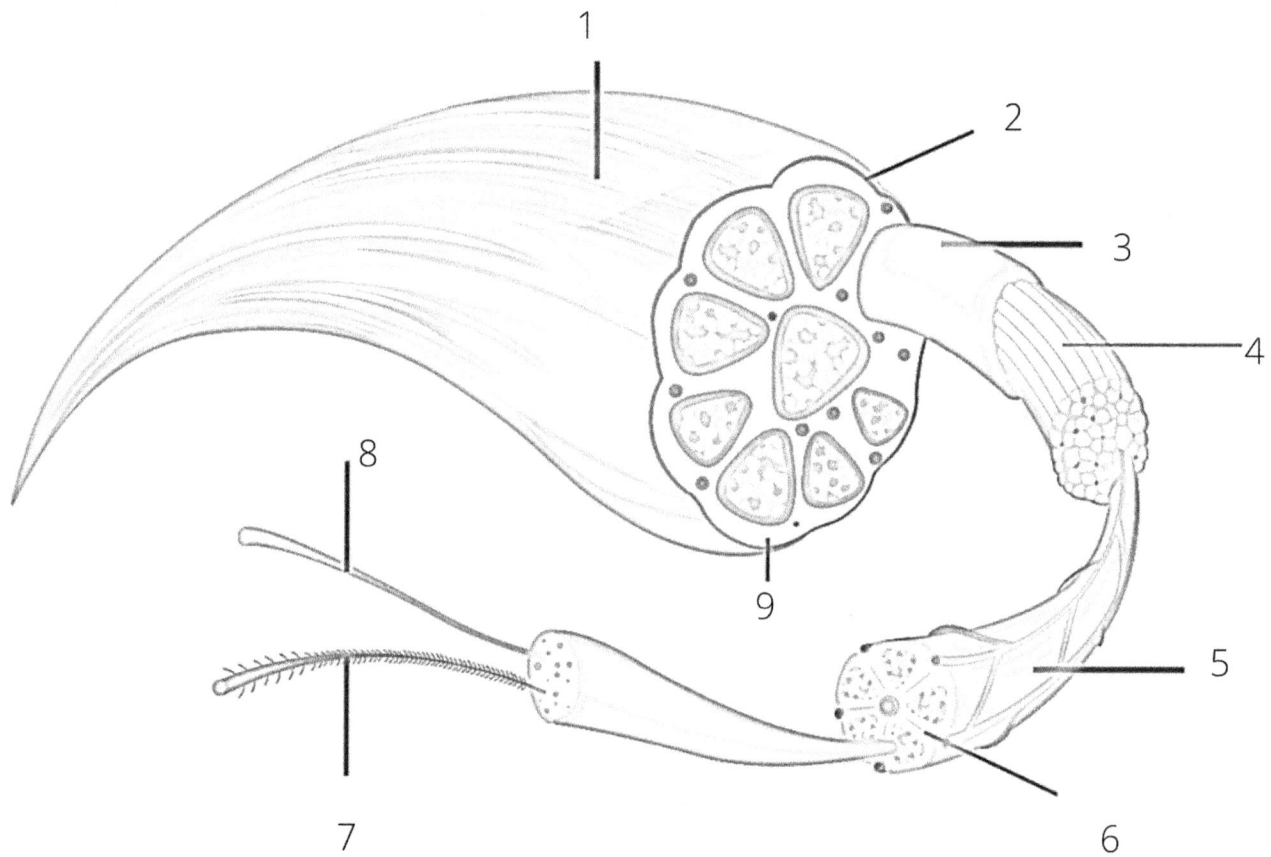

Type of Muscle Cells

Select different colors for each of the areas provided with a coding numbers and corresponding structures in the diagram.

1. Cardiac Muscle
2. Skeletal Muscle
3. Smooth Muscle

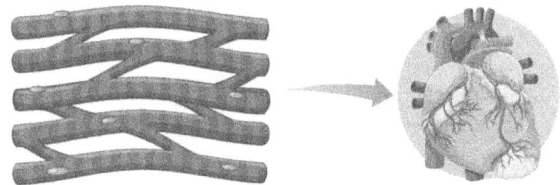

Cardiac muscle

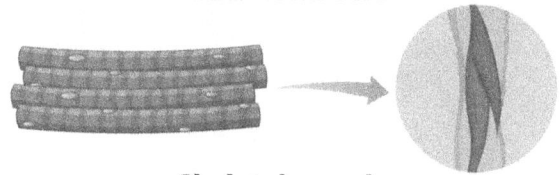

Skeletal muscle

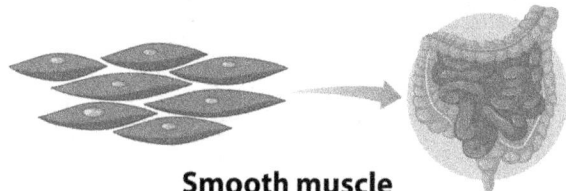

Smooth muscle

_____

_____

_____

_____

_____

_____

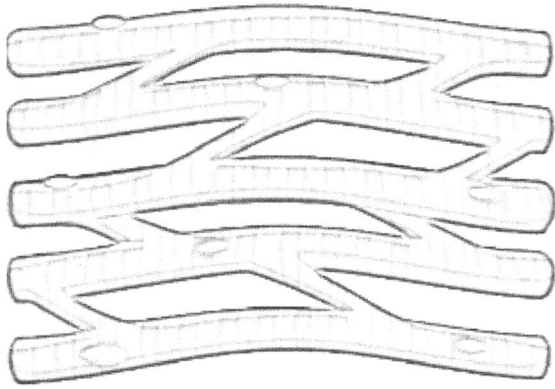  

1 _____

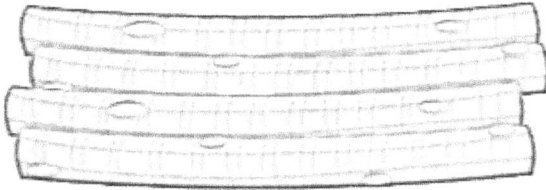  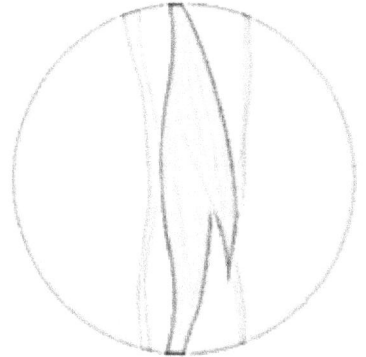

2 _____

  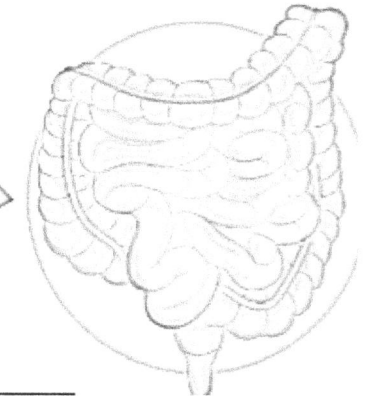

3 _____

Leg Muscle Anatomy

Select different colors for each of the areas provided with a coding numbers and corresponding structures in the diagram.

1. Sartorius
2. Abductor Longus
3. Patella
4. Gastrocnemius
5. Soleus
6. Peroneus Brevis
7. Tibialis Anterior
8. Peroneus Longus
9. Vastus Lateralis
10. Rectus Femoris
11. Tensor Fasciae Latae
12. Anterior Iliac Spine

Leg Muscle Anatomy (Front View)

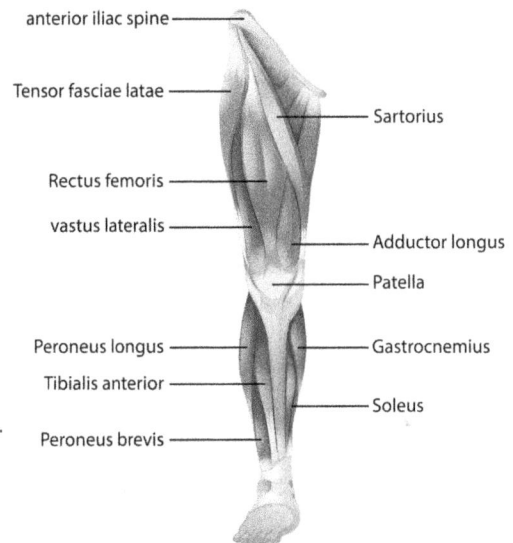

anterior iliac spine
Tensor fasciae latae
Sartorius
Rectus femoris
vastus lateralis
Adductor longus
Patella
Peroneus longus
Gastrocnemius
Tibialis anterior
Soleus
Peroneus brevis

_____

_____

_____

_____

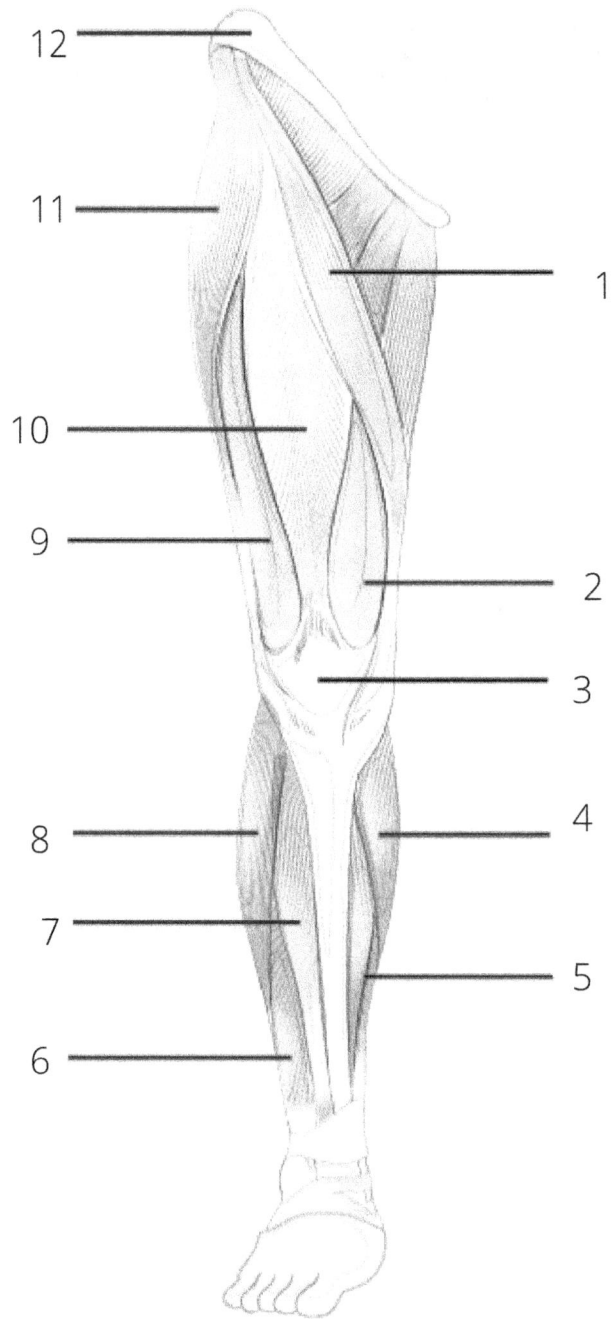

How Muscles Work

Select different colors for each of the areas provided with a coding numbers and corresponding structures in the diagram.

1. Biceps Relaxed
2. Triceps Contracted
3. Biceps Contracted
4. Triceps Relaxed

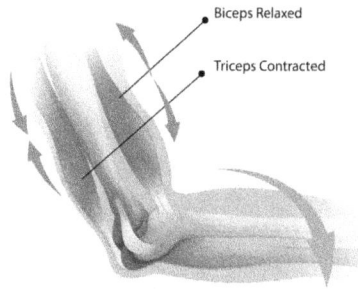

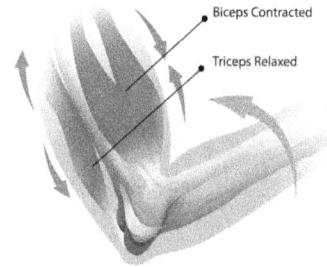

Biceps Relaxed
Triceps Contracted
Biceps Contracted
Triceps Relaxed

_____

_____

_____

_____

_____

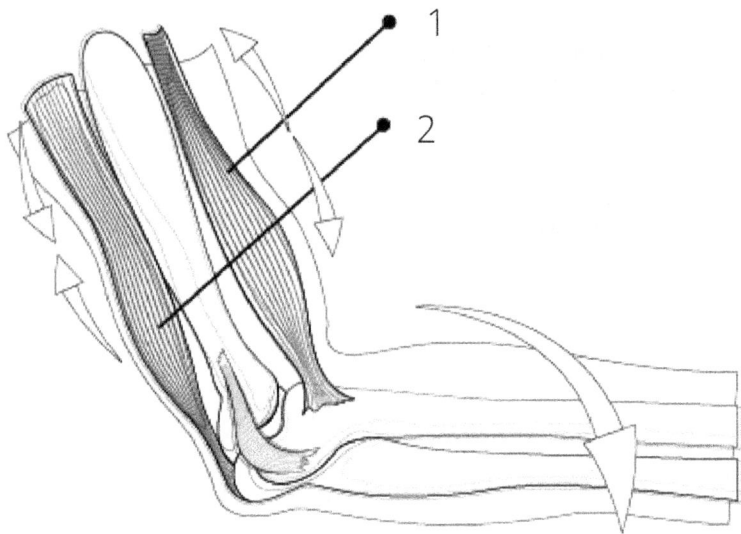

1

2

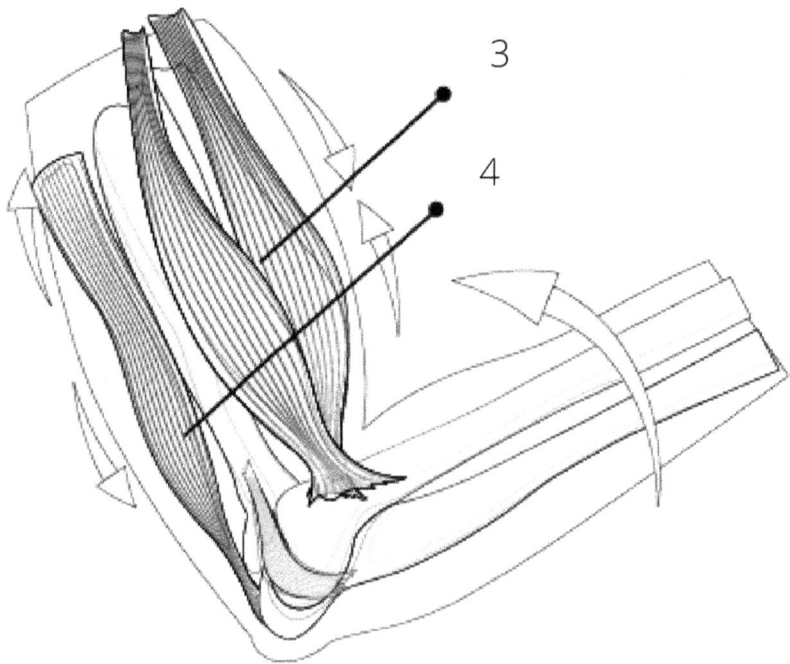

3

4

7. Nervous System

Features of the Nervous System:

- To find changes and feel sensitivities.
- To initiate appropriate responses to changes.
- To arrange information for immediate use and store it for future use.

The nervous system has two divisions. The central nervous system (CNS) contains the human brain and spinal cord. The peripheral nervous system (PNS) consists of cranial and spinal nerves. In addition, the PNS includes the Automatic nerve system (ANS).

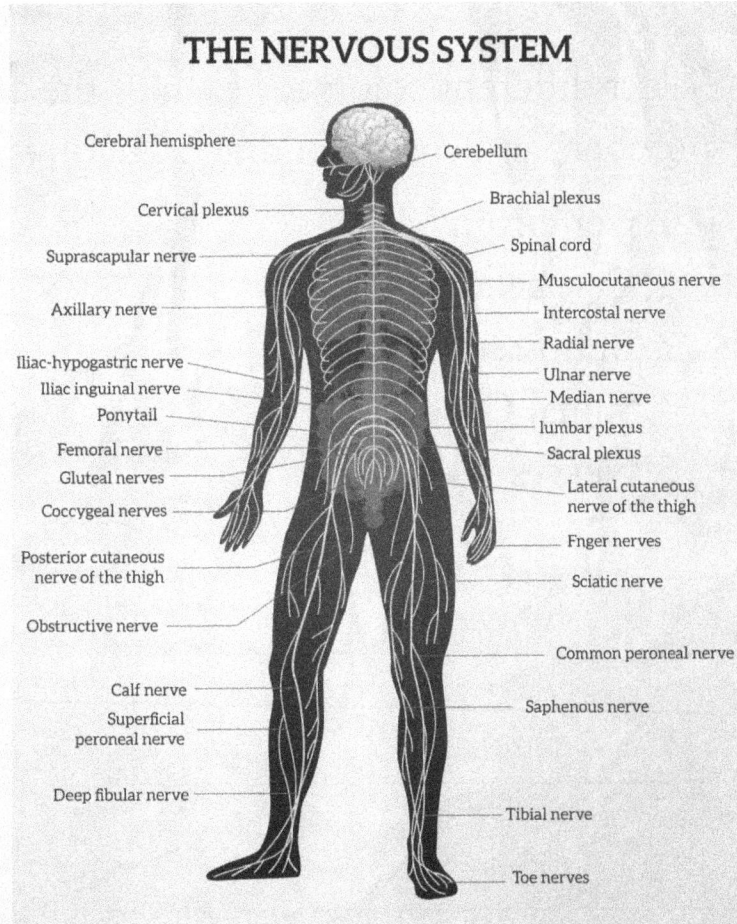

THE NERVOUS SYSTEM

Cerebral hemisphere
Cerebellum
Cervical plexus
Brachial plexus
Suprascapular nerve
Spinal cord
Musculocutaneous nerve
Axillary nerve
Intercostal nerve
Radial nerve
Iliac-hypogastric nerve
Ulnar nerve
Iliac inguinal nerve
Median nerve
Ponytail
lumbar plexus
Femoral nerve
Sacral plexus
Gluteal nerves
Lateral cutaneous nerve of the thigh
Coccygeal nerves
Fnger nerves
Posterior cutaneous nerve of the thigh
Sciatic nerve
Obstructive nerve
Common peroneal nerve
Calf nerve
Saphenous nerve
Superficial peroneal nerve
Deep fibular nerve
Tibial nerve
Toe nerves

Nervous System

Select different colors for each of the areas provided with a coding numbers and corresponding structures in the diagram.

1. Cerebrum
2. Cerebellum
3. Brainstem
4. Spinal Cord
5. Nerves

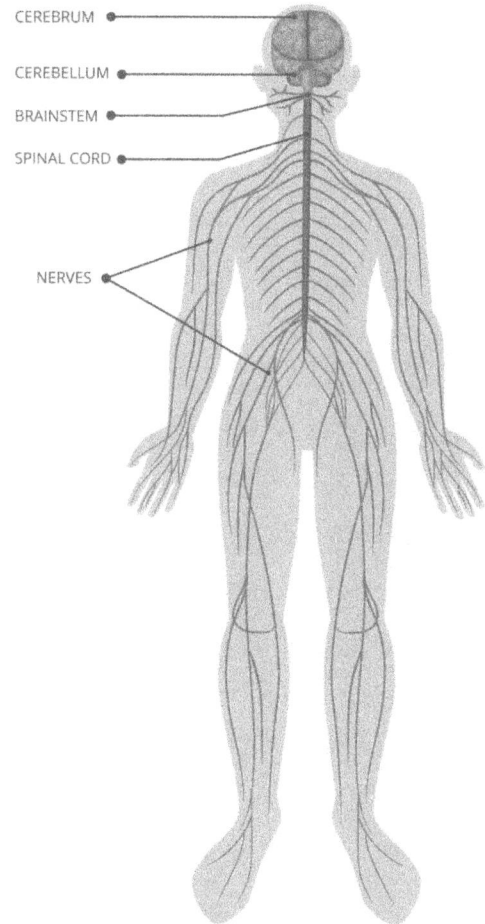

CEREBRUM

CEREBELLUM

BRAINSTEM

SPINAL CORD

NERVES

_____

_____

_____

_____

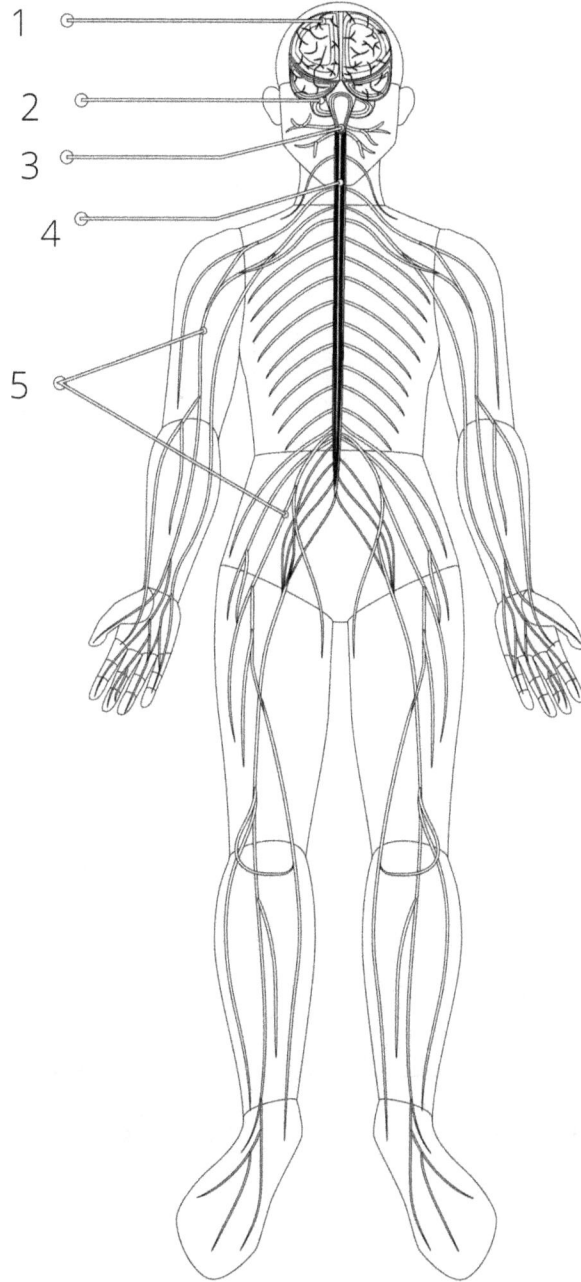

1

2

3

4

5

5

Neuron Anatomy

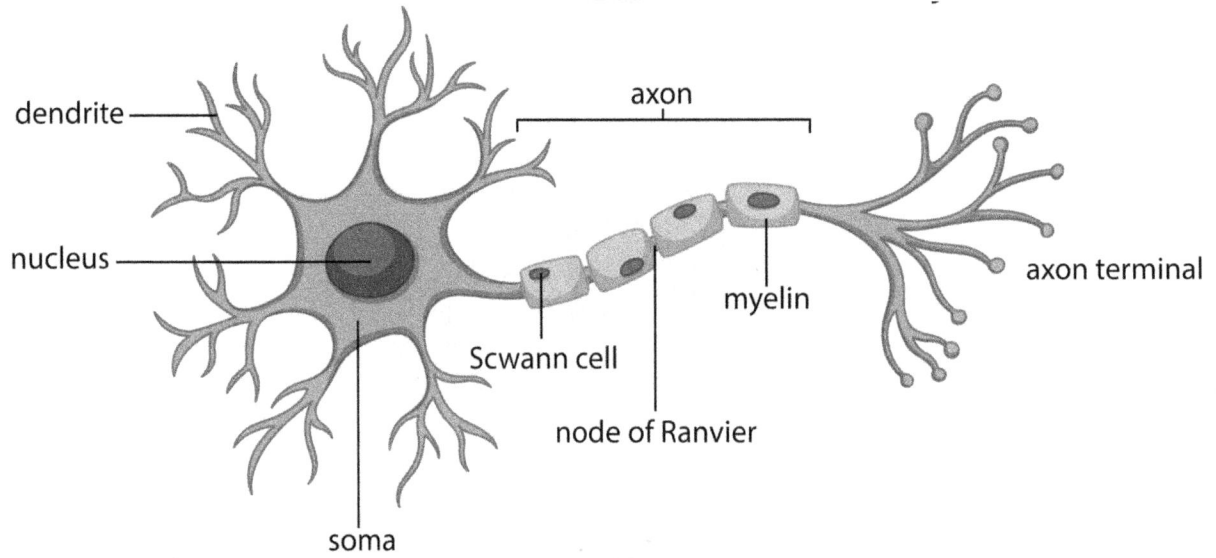

dendrite

axon

nucleus

axon terminal

Scwann cell

myelin

node of Ranvier

soma

The neuron is the standard working device of the brain, a specialized cell made to send details to another neuron, muscle mass, or gland cells. Neurons are cells within the nerves that transfer info to other nerve cells, muscular tissue, or gland cells. Many neurons have a cell body, an axon, and also dendrites.

Neuron Anatomy

Select different colors for each of the areas provided with a coding numbers and corresponding structures in the diagram.

1. Dendrite
2. Nucleus
3. Soma
4. Scwann Cell
5. Node of Ranvier

6. Myelin
7. Axon Terminal
8. Axon

_____

_____

_____

_____

Coloring Page- Neuron

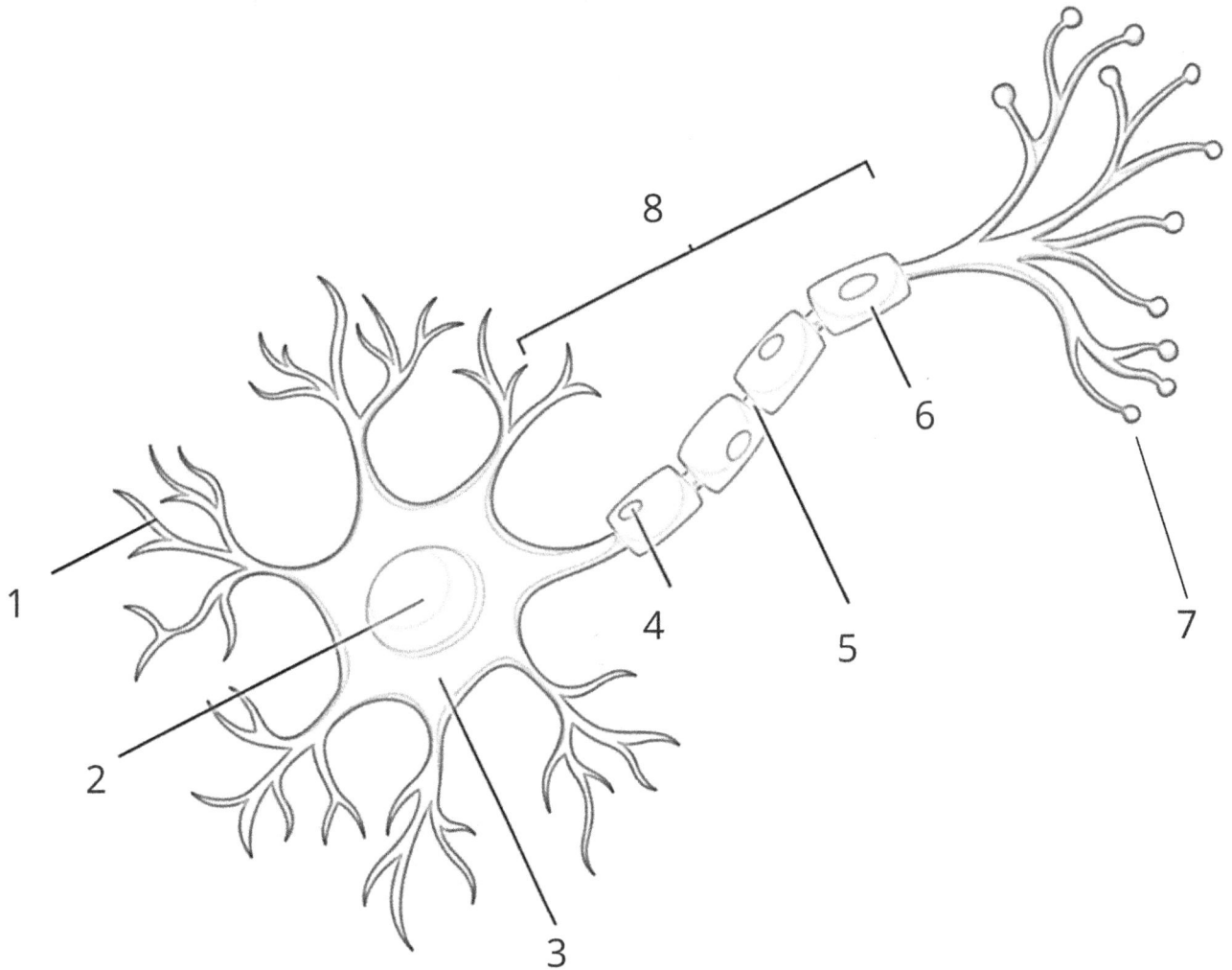

1

2

3

4

5

6

7

8

Human Brain

Select different colors for each of the areas provided with a coding numbers and corresponding structures in the diagram.

1. Frontal Lobe
2. Temporal Lobe
3. Spinal Cord
4. Cerebellum
5. Occipital Lobe
6. Parietal Lobe

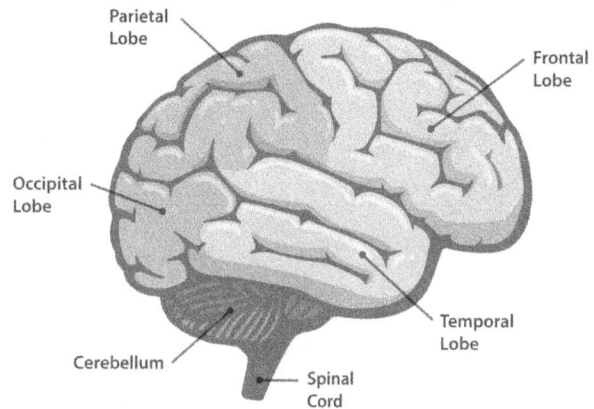

Parietal Lobe

Frontal Lobe

Occipital Lobe

Temporal Lobe

Cerebellum

Spinal Cord

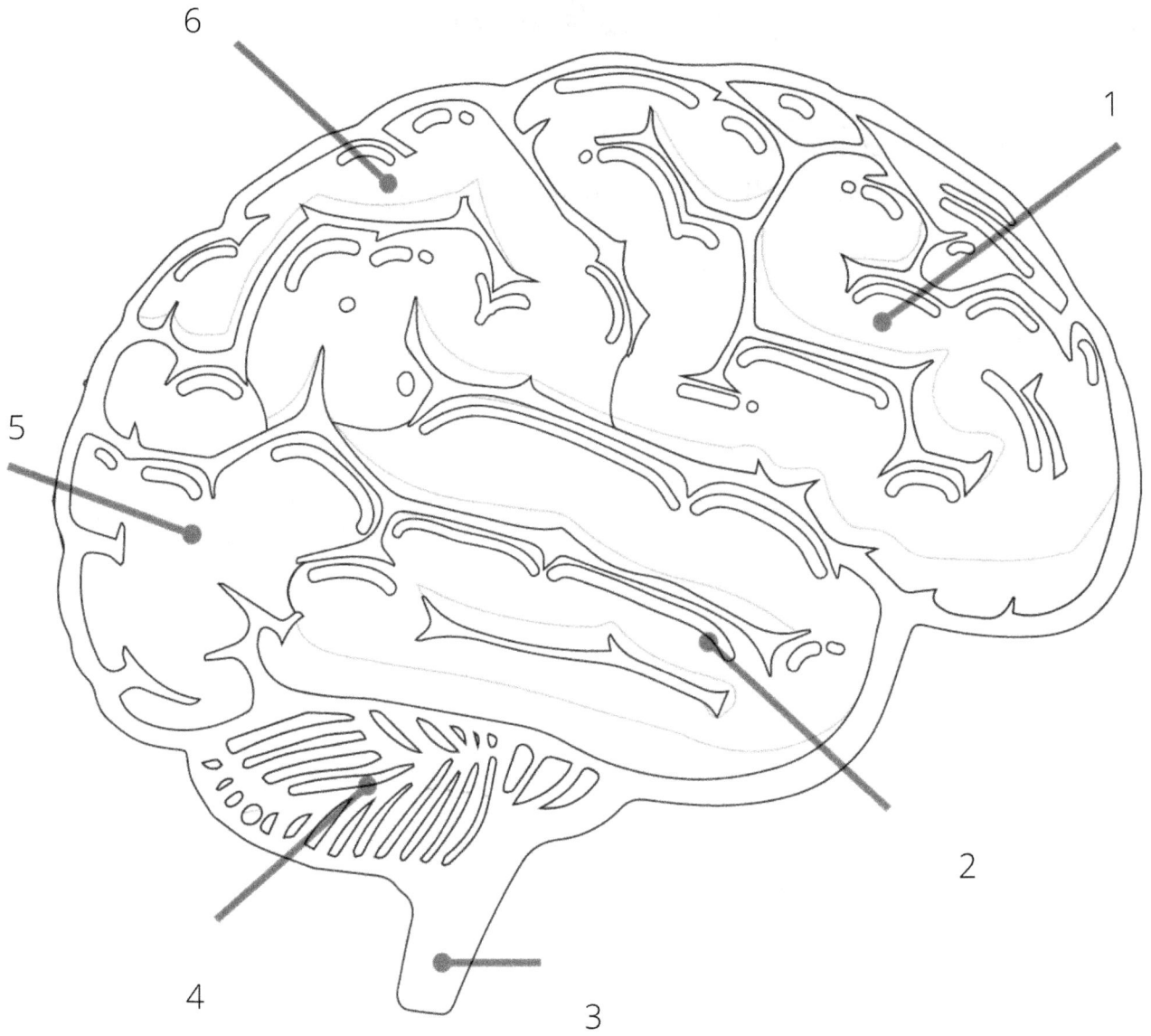

Eye Anatomy

Select different colors for each of the areas provided with a coding numbers and corresponding structures in the diagram.

1. Superior Rectus Muscle
2. Retina
3. Optic Nerve
4. Vitreous Gel
5. Inferior Rectus Muscle
6. Choroid
7. Lens
8. Cornea
9. Pupil
10. Anterior Chamber
11. Iris

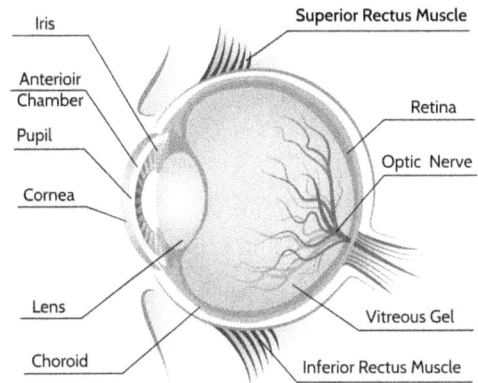

_____

_____

_____

_____

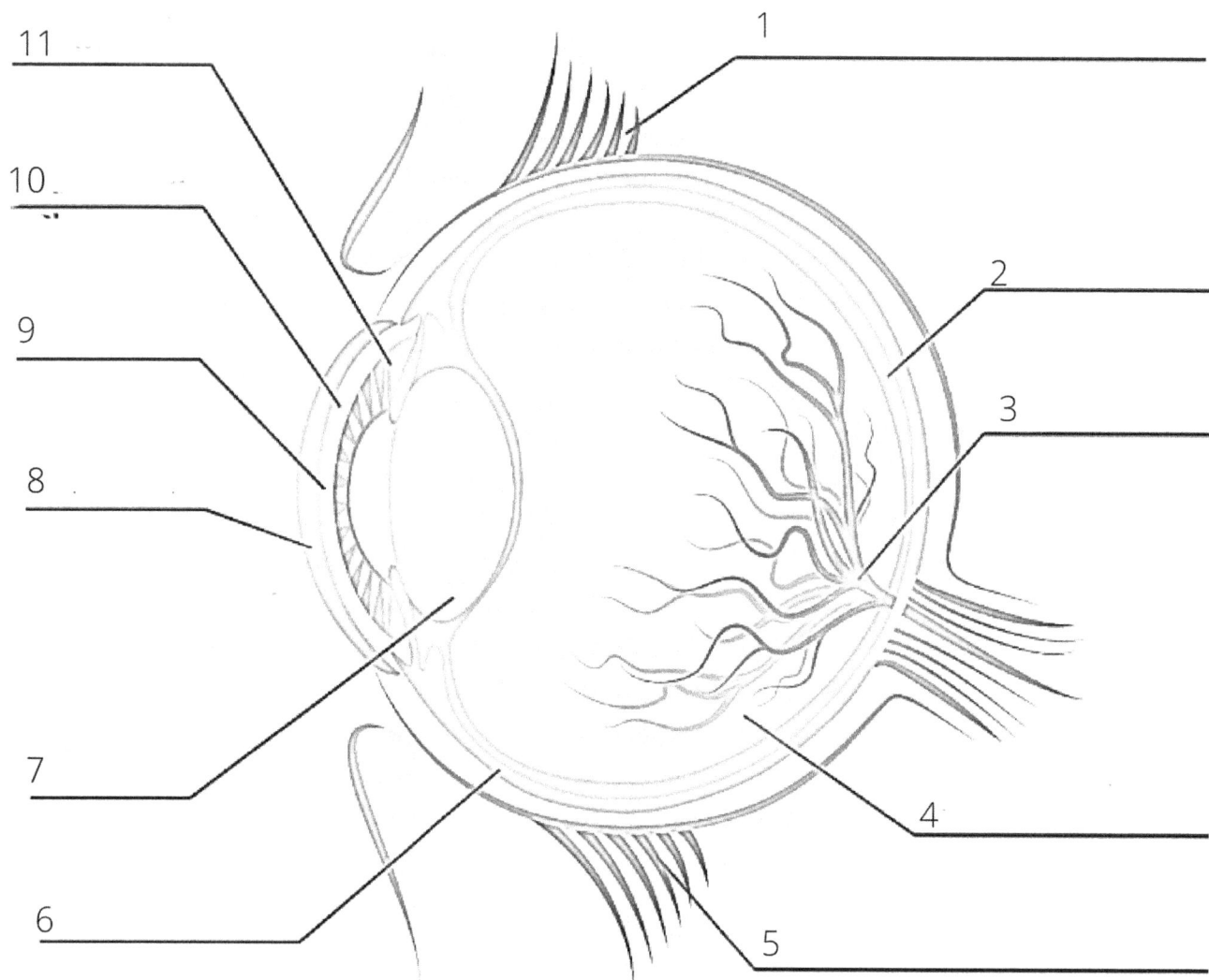

11

10

9

8

7

6

1

2

3

4

5

8.Endocrine System

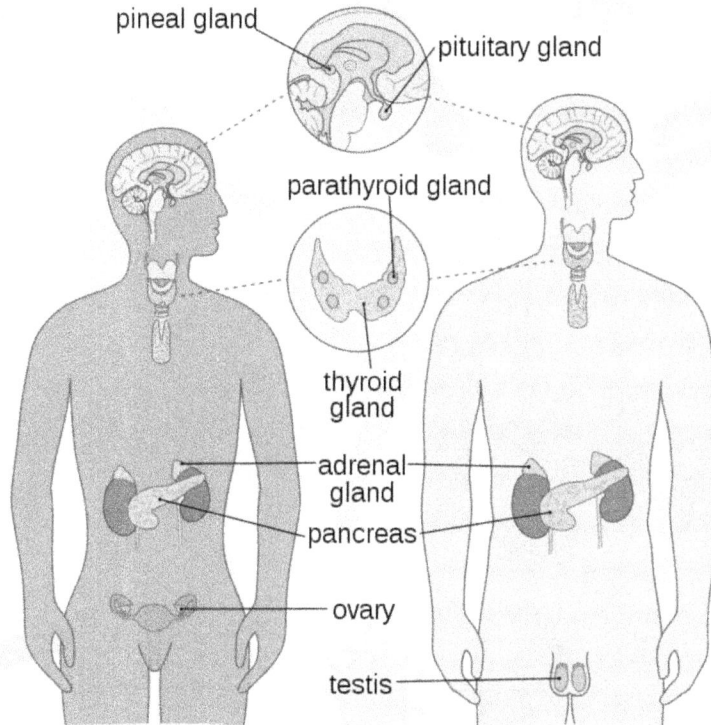

pineal gland

pituitary gland

parathyroid gland

thyroid gland

adrenal gland

pancreas

ovary

testis

The other controlling system of the body is the endocrine system, which contains endocrine glands that produce chemicals called hormones.

It contains the hypothalamus, pituitary, thyroid, parathyroid, pineal, thymus glands, pancreatic, ovaries, testes, and other organs that produce hormones.

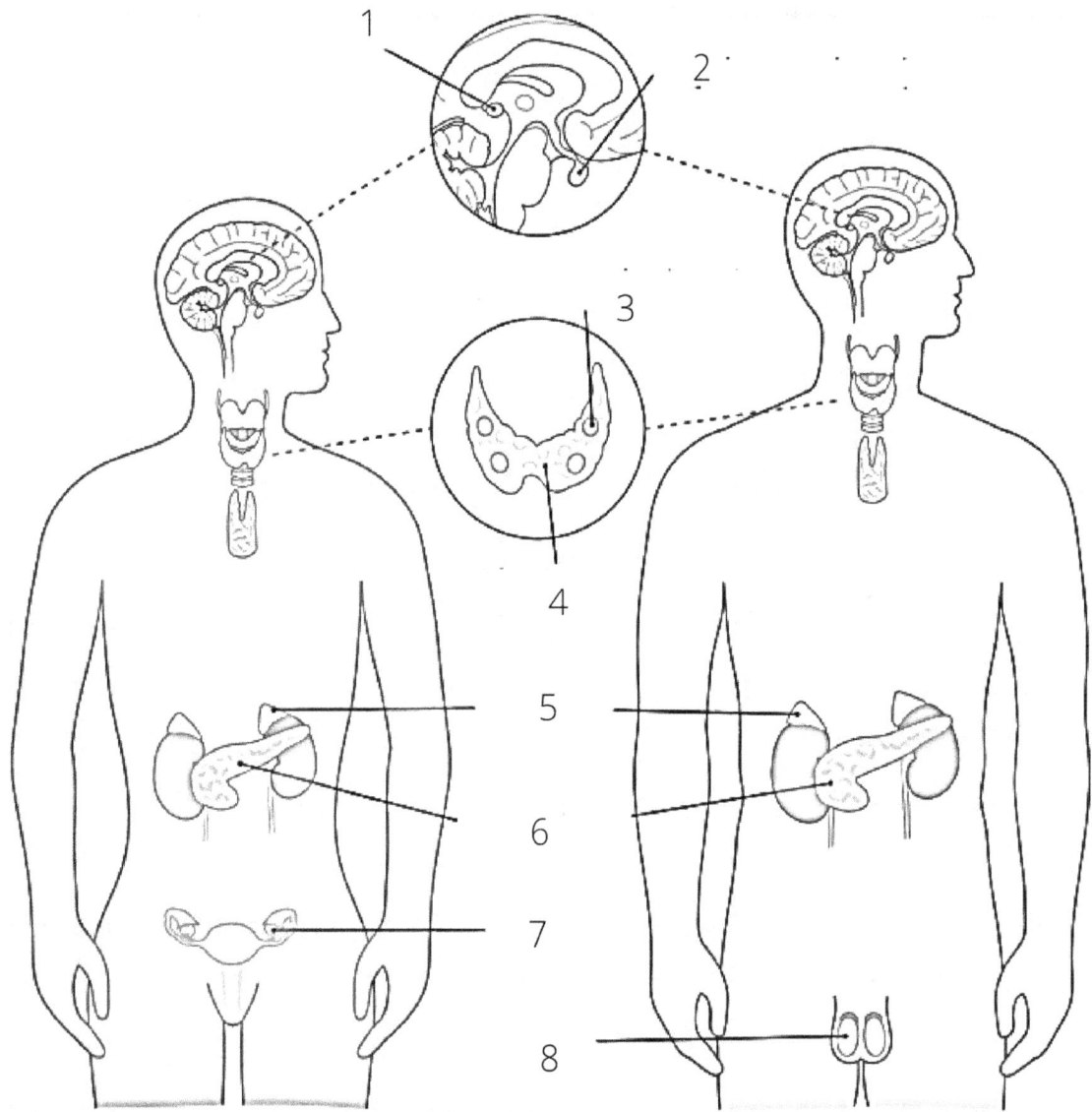

Pineal Gland

Select different colors for each of the areas provided with a coding numbers and corresponding structures in the diagram.

1. Pineal Gland
2. Melatonin

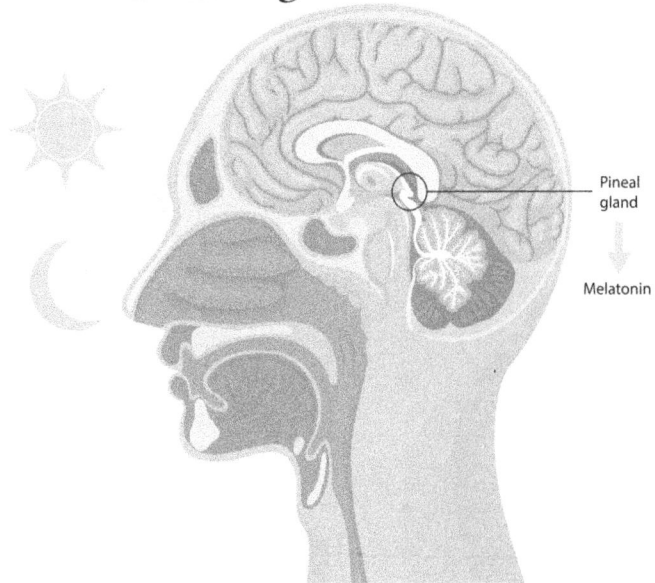

Pineal gland

Melatonin

_____

_____

_____

_____

_____

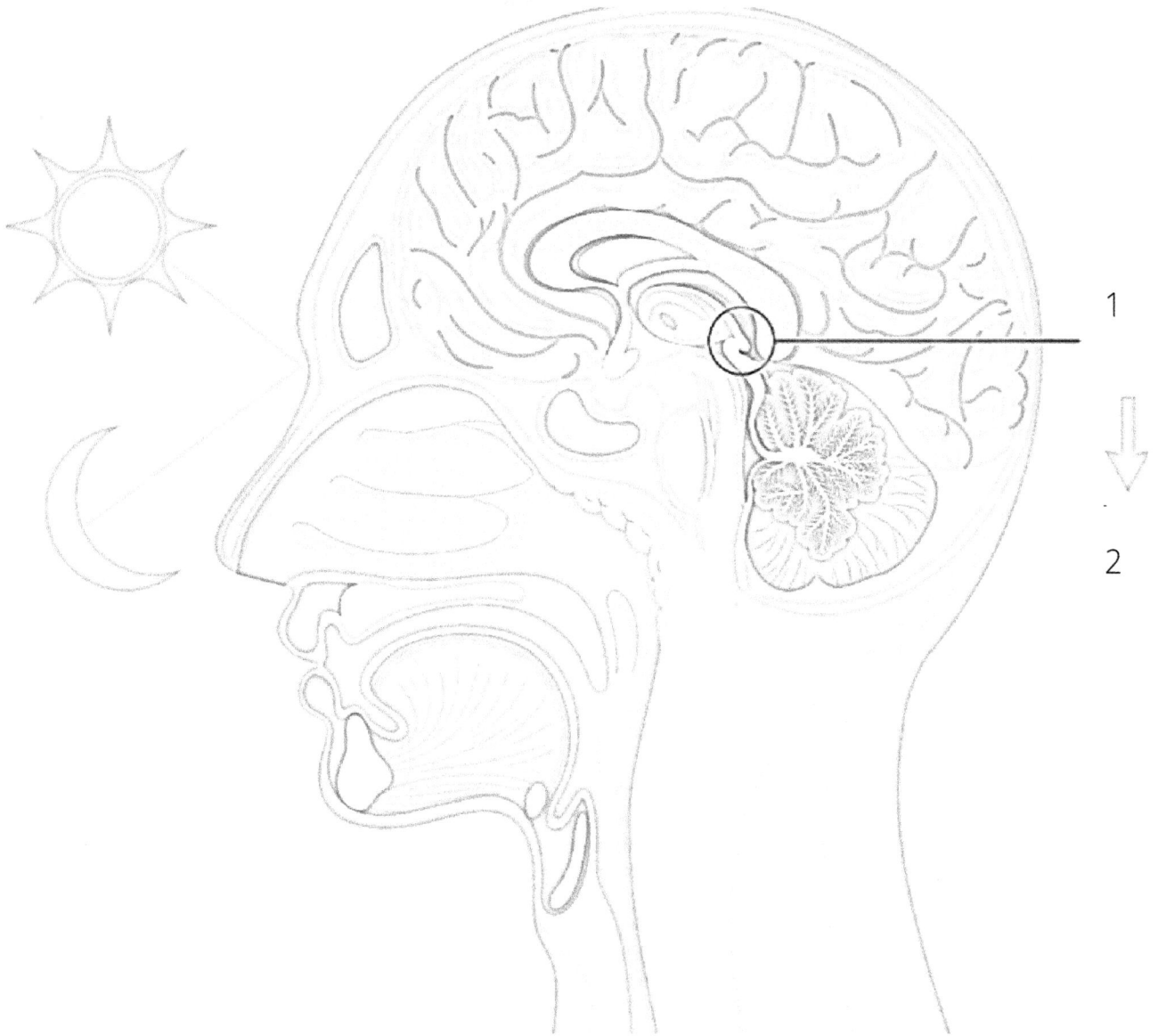

1

2

Pituitary Gland

Select different colors for each of the areas provided with a coding numbers and corresponding structures in the diagram.

1. Optic Chiasm
2. Anterior Pituitary
3. Posterior Pituitary
4. Hypothalamus

PITUITARY GLAND

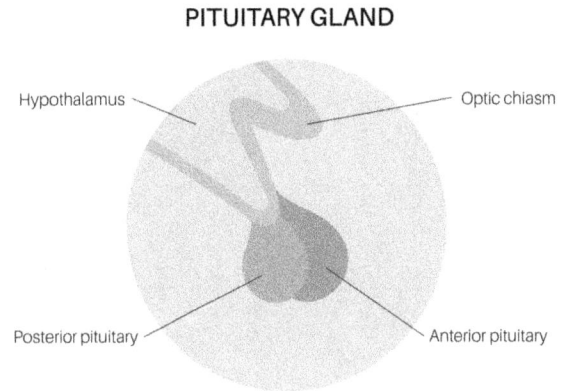

Hypothalamus

Optic chiasm

Posterior pituitary

Anterior pituitary

_____

_____

_____

_____

_____

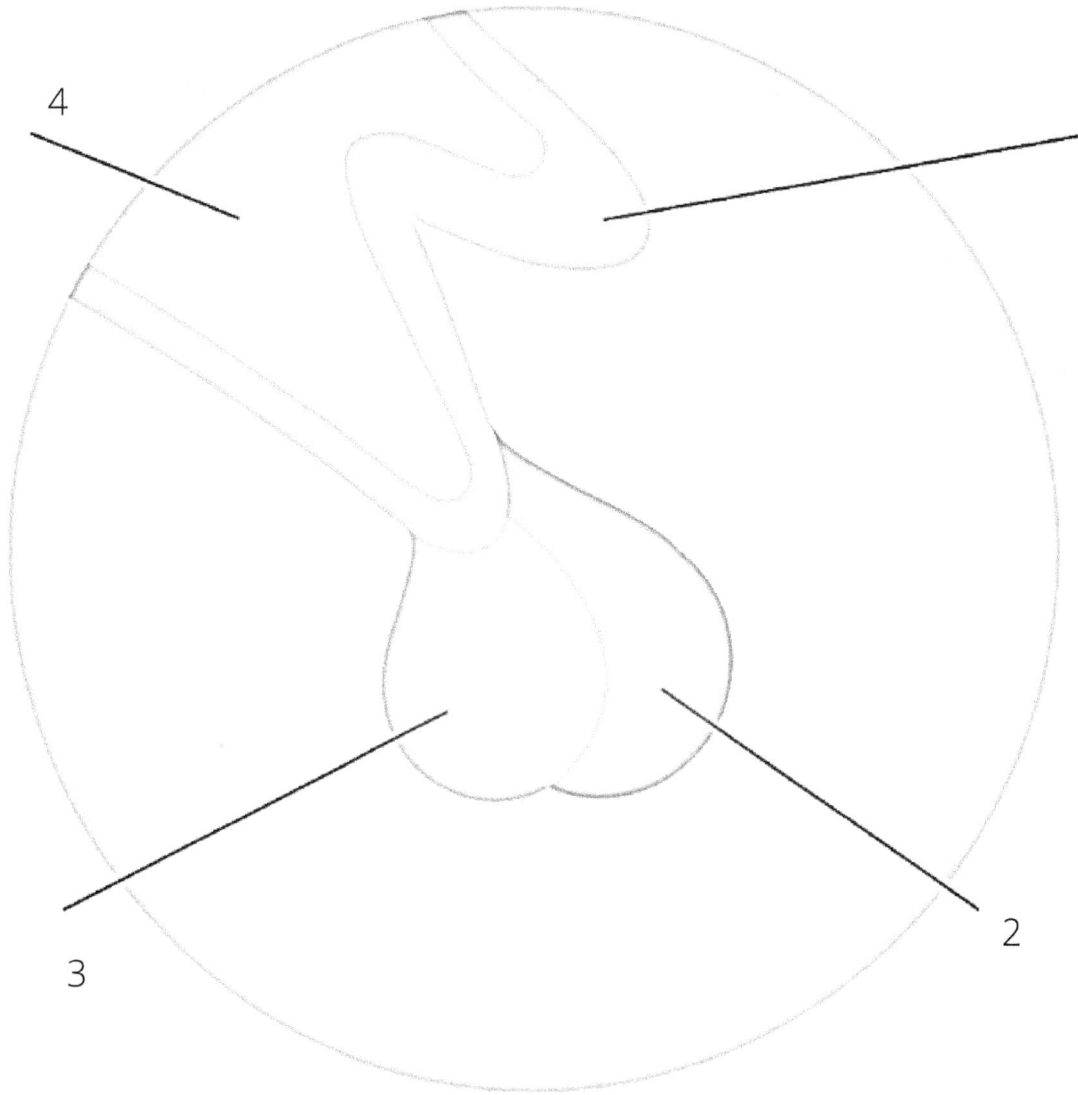

Thyroid System

Select different colors for each of the areas provided with a coding numbers and corresponding structures in the diagram.

1. Hypothalamus
2. Pituitary Gland
3. Thyroid Gland
4. Triiodothyronine
5. Thyroxine
6. Calcitonin

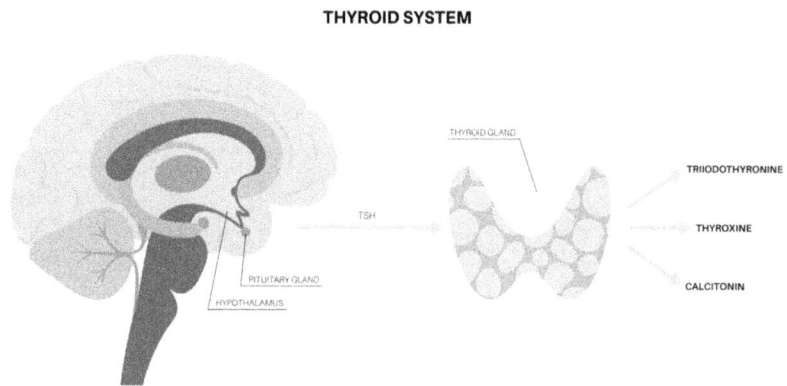

THYROID SYSTEM

THYROID GLAND

TSH

TRIIODOTHYRONINE

THYROXINE

CALCITONIN

PITUITARY GLAND

HYPOTHALAMUS

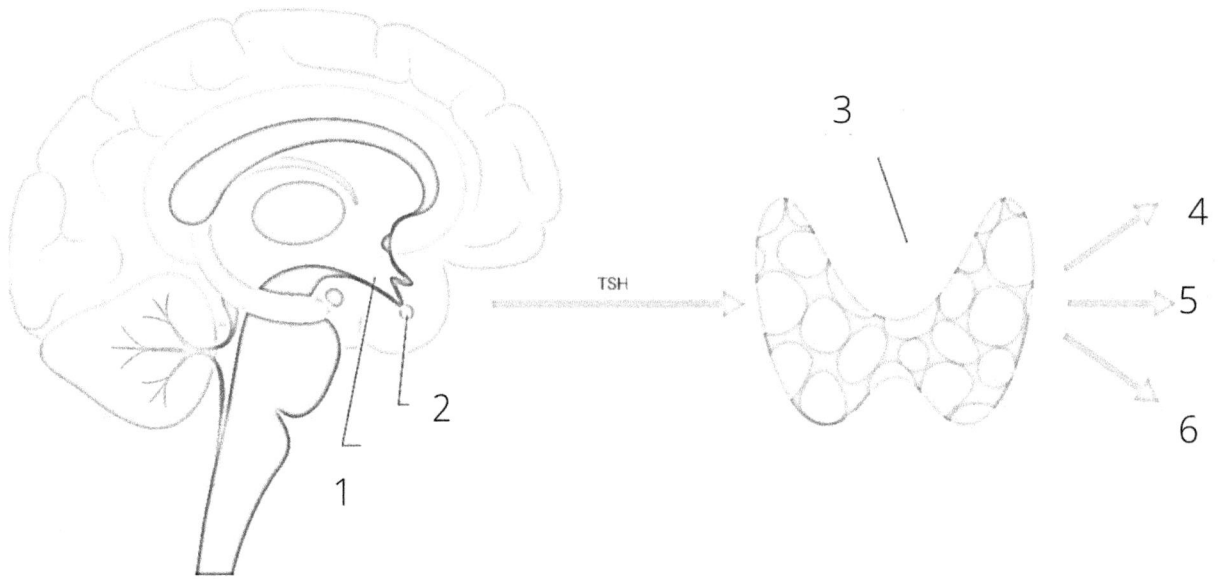

TSH

1

2

3

4

5

6

Thyroid Hormone

Select different colors for each of the areas provided with a coding numbers and corresponding structures in the diagram.

1. Hypothalamus
2. Pituitary Gland
3. Thyroid Gland
4. Calcitonin

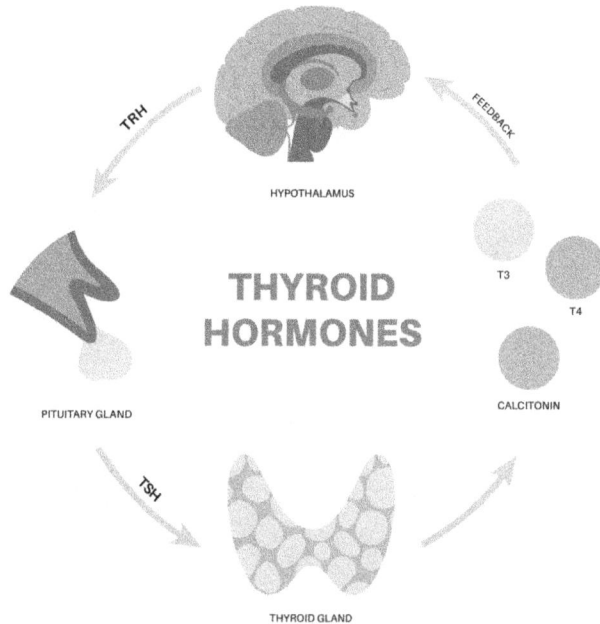

TRH

FEEDBACK

HYPOTHALAMUS

THYROID HORMONES

T3

T4

CALCITONIN

PITUITARY GLAND

TSH

THYROID GLAND

_____

_____

_____

_____

_____

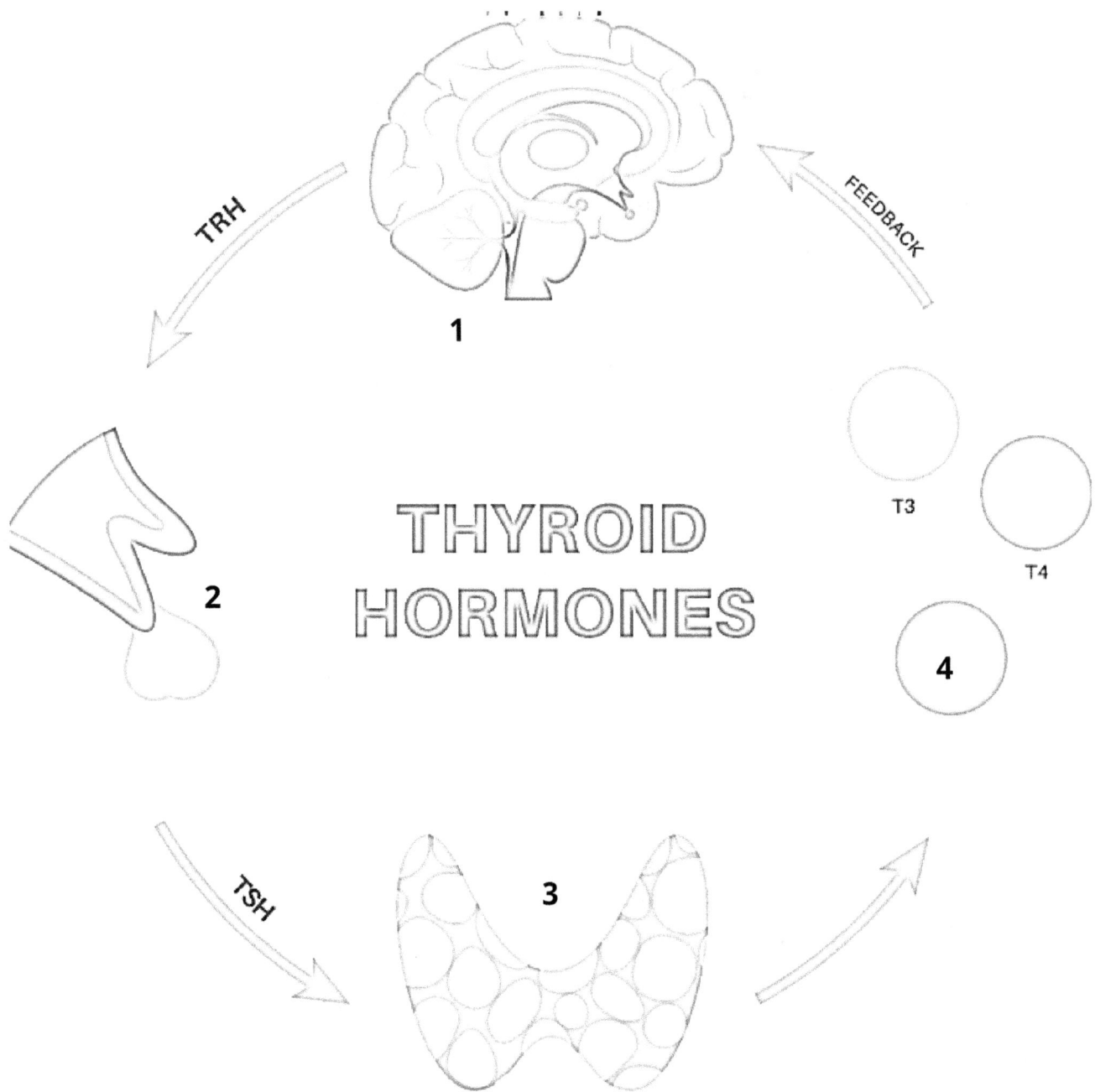

THYROID HORMONES

1

2

3

4

T3

T4

TRH

TSH

FEEDBACK

9.Circulatory System

The blood circulatory system consistes of organs that includes the heart and blood vessels, as well as blood circulates throughout a human or vertebrate's entire body.

In addition, it has the cardiovascular or vascular system, which includes the heart and blood vessels.

The circulatory system consists of four significant components:
- Heart.
- Arteries.
- Veins.
- Blood.

The circulatory system contains three independent systems that interact: the heart (cardiovascular), lungs (pulmonary), arteries, blood vessels, and coronary and also portal vessels (systemic). The system is in charge of the flow of blood, nutrients, oxygen, as well as various other gases and along with hormones to and from cells.

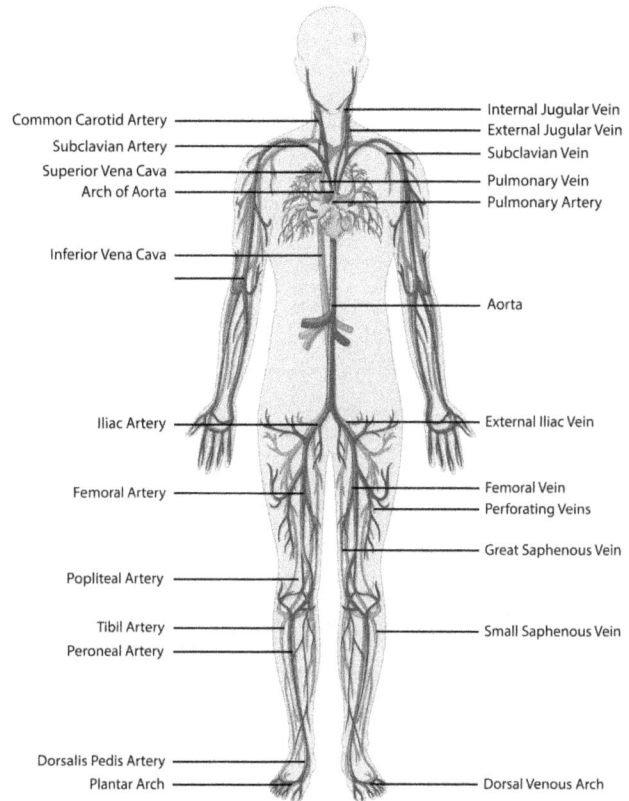

Common Carotid Artery
Subclavian Artery
Superior Vena Cava
Arch of Aorta
Inferior Vena Cava
Iliac Artery
Femoral Artery
Popliteal Artery
Tibil Artery
Peroneal Artery
Dorsalis Pedis Artery
Plantar Arch

Internal Jugular Vein
External Jugular Vein
Subclavian Vein
Pulmonary Vein
Pulmonary Artery
Aorta
External Iliac Vein
Femoral Vein
Perforating Veins
Great Saphenous Vein
Small Saphenous Vein
Dorsal Venous Arch

Human Circulatory System

Select different colors for each of the areas provided with a coding numbers and corresponding structures in the diagram.

1. Pulmonary Artery
2. Aorta
3. Vein
4. Anterior vena cava

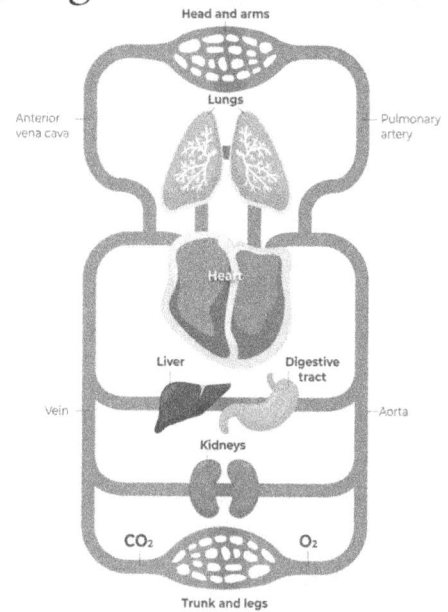

_____

_____

_____

_____

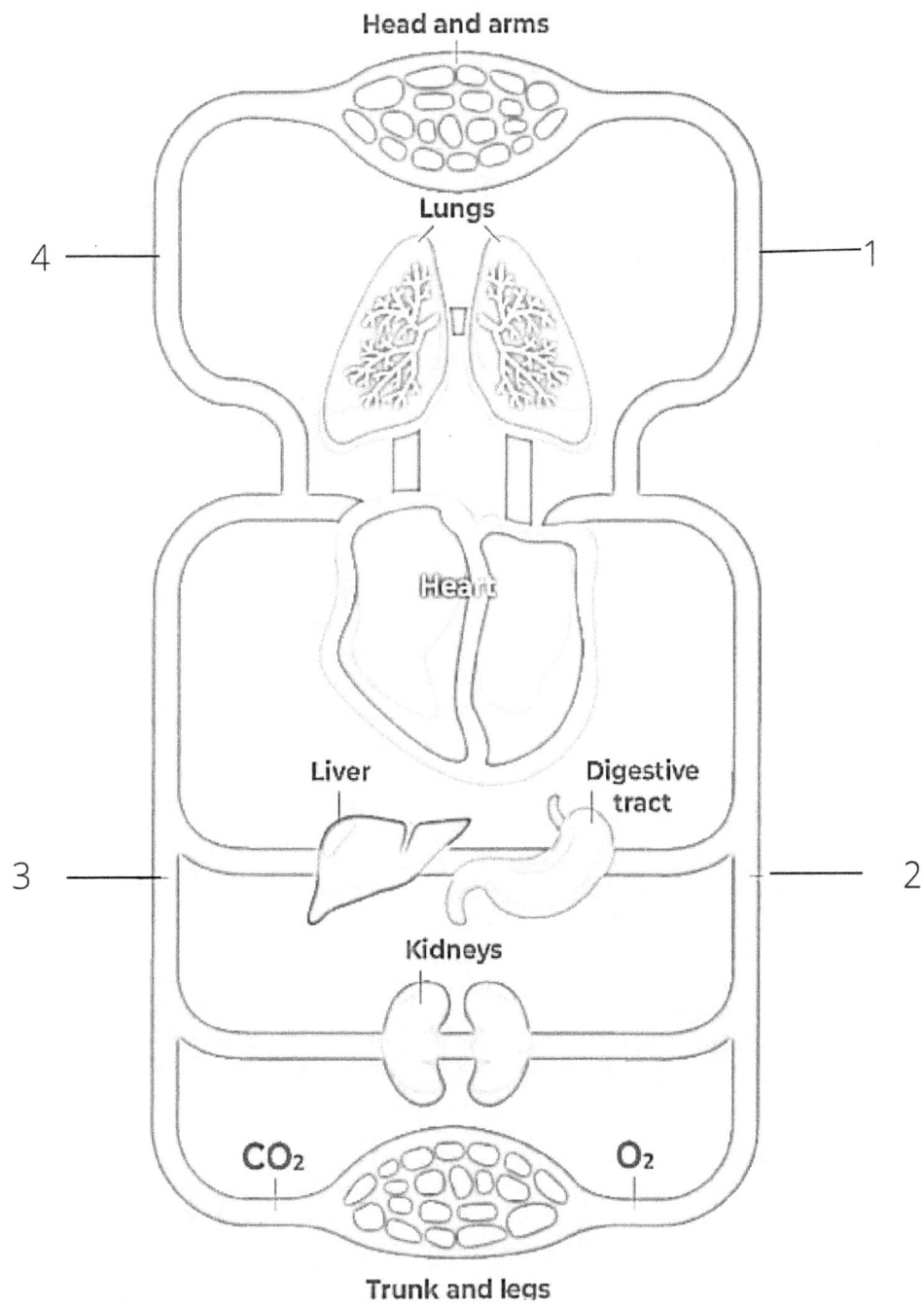

Head and arms

4

1

Lungs

Heart

Liver

Digestive tract

3

2

Kidneys

CO_2

O_2

Trunk and legs

Heart

Select different colors for each of the areas provided with a coding numbers and corresponding structures in the diagram.

1. Aorta
2. Pulmonary Artery
3. Pulmonary Veins
4. Auricle of Left Atrium
5. Left Coronary Artery
6. Left Ventricle
7. Inferior Venacava
8. Right Atrium
9. Right Pulmonary Veins
10. Superior Venacava

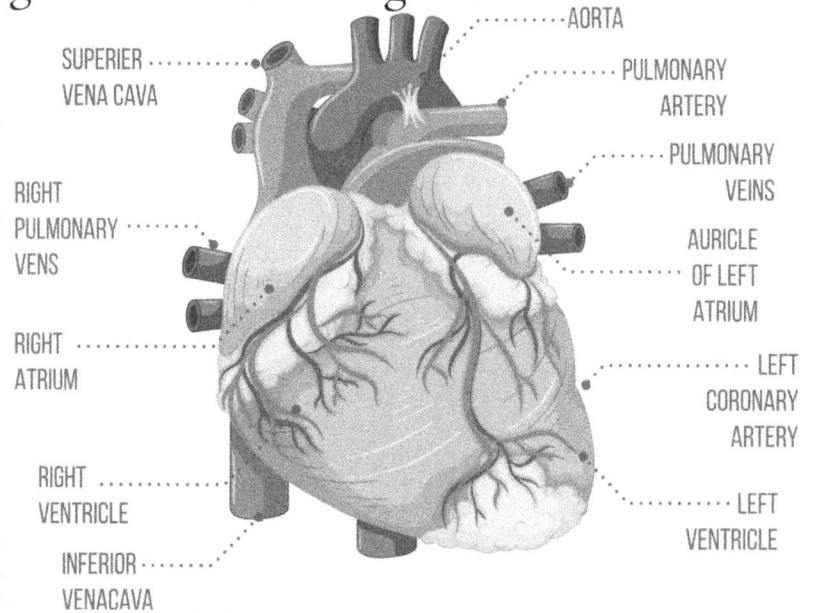

SUPERIER VENA CAVA

RIGHT PULMONARY VENS

RIGHT ATRIUM

RIGHT VENTRICLE

INFERIOR VENACAVA

AORTA

PULMONARY ARTERY

PULMONARY VEINS

AURICLE OF LEFT ATRIUM

LEFT CORONARY ARTERY

LEFT VENTRICLE

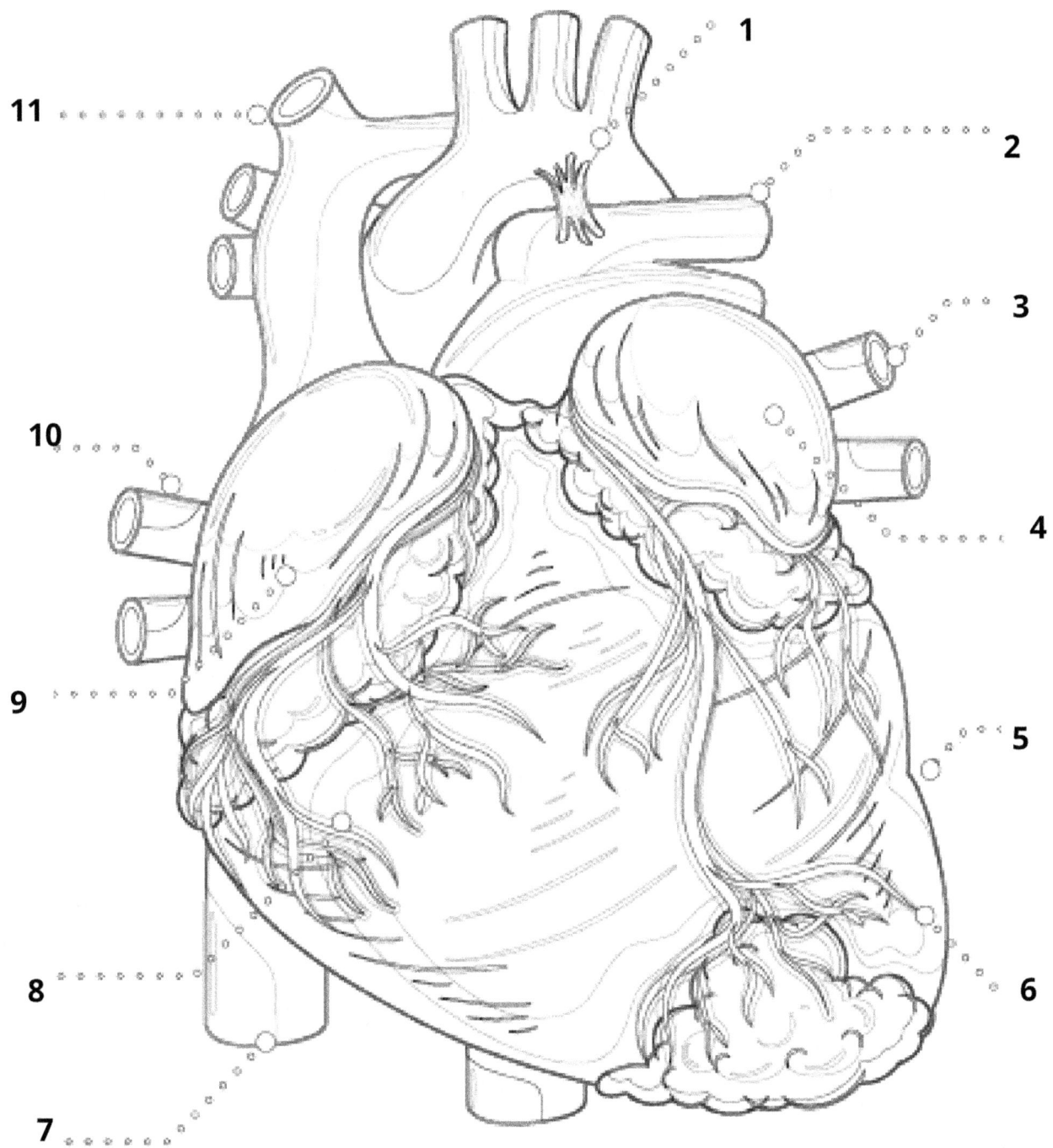

Cardiovascular System

Select different colors for each of the areas provided with a coding numbers and corresponding structures in the diagram.

1. Elastic Artery
2. Muscular Artery
3. Arteriole
4. Capillaries
5. Venola
6. Middle Vein
7. Greater Vienna

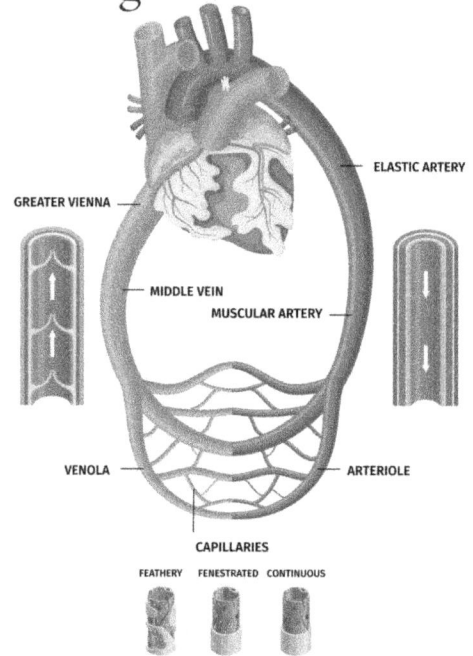

ELASTIC ARTERY

GREATER VIENNA

MIDDLE VEIN

MUSCULAR ARTERY

VENOLA

ARTERIOLE

CAPILLARIES

FEATHERY   FENESTRATED   CONTINUOUS

_____

_____

_____

_____

1

7

6

2

5

3

4

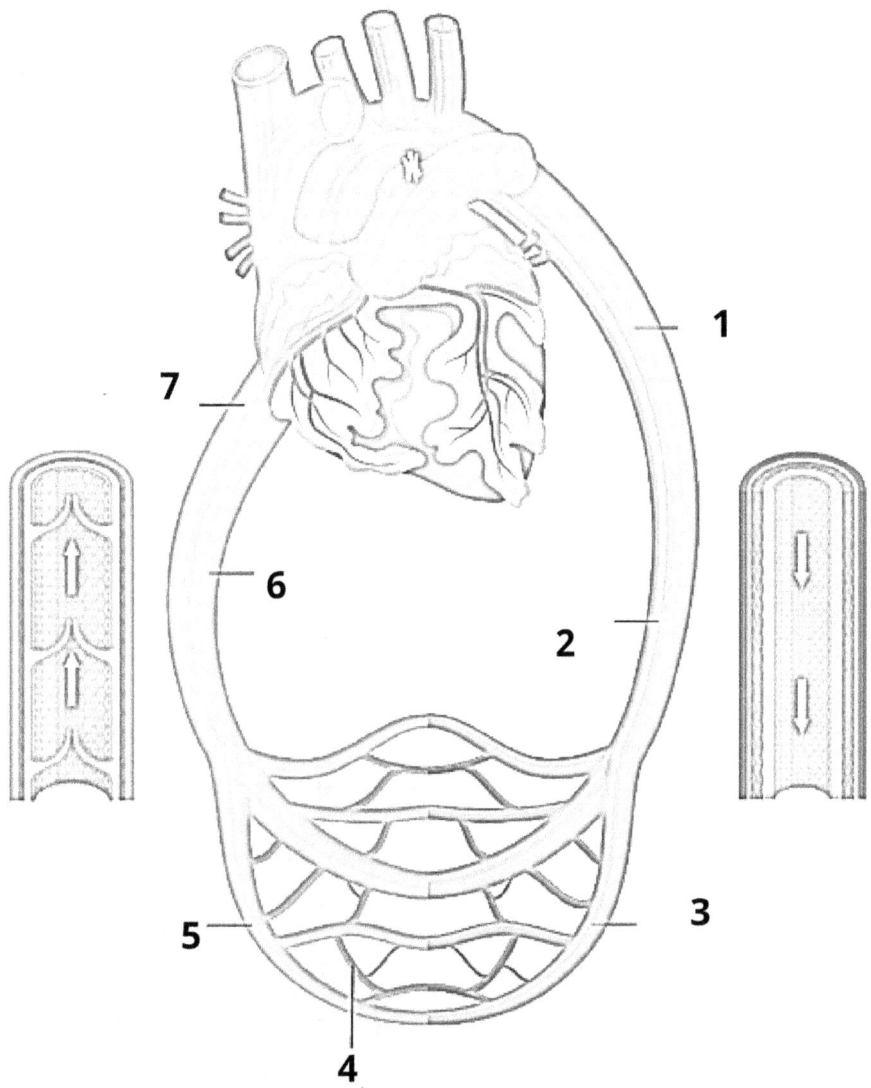

FEATHERY   FENESTRATED   CONTINUOUS

10.Lymphatic System

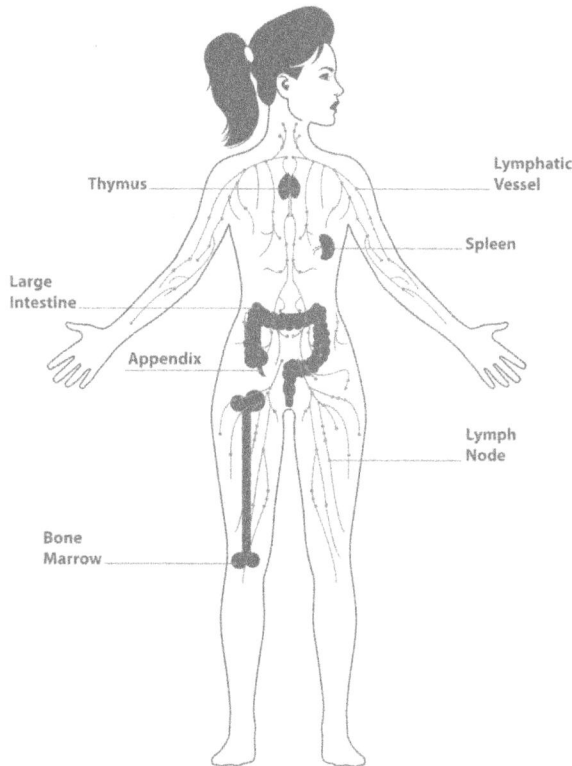

The lymphatic system includes lymphatic vessels, lymph nodes, the thymus, and the spleen. It drains excess tissue fluid and cells of immunity.

The lymphatic system has three features:
- The removal of excess fluids from body tissues.
- Absorption of fatty acids and subsequent transportation of fat, chyle, to the blood circulation system.
- Manufacturing of immune cells

Thymus

Large Intestine

Appendix

Bone Marrow

Lymphatic Vessel

Spleen

Lymph Node

1. Lymphatic Vessel
2. Spleen
3. Lymph Node
4. Bone Marrow
5. Appendix
6. Large Intestine
7. Thymus

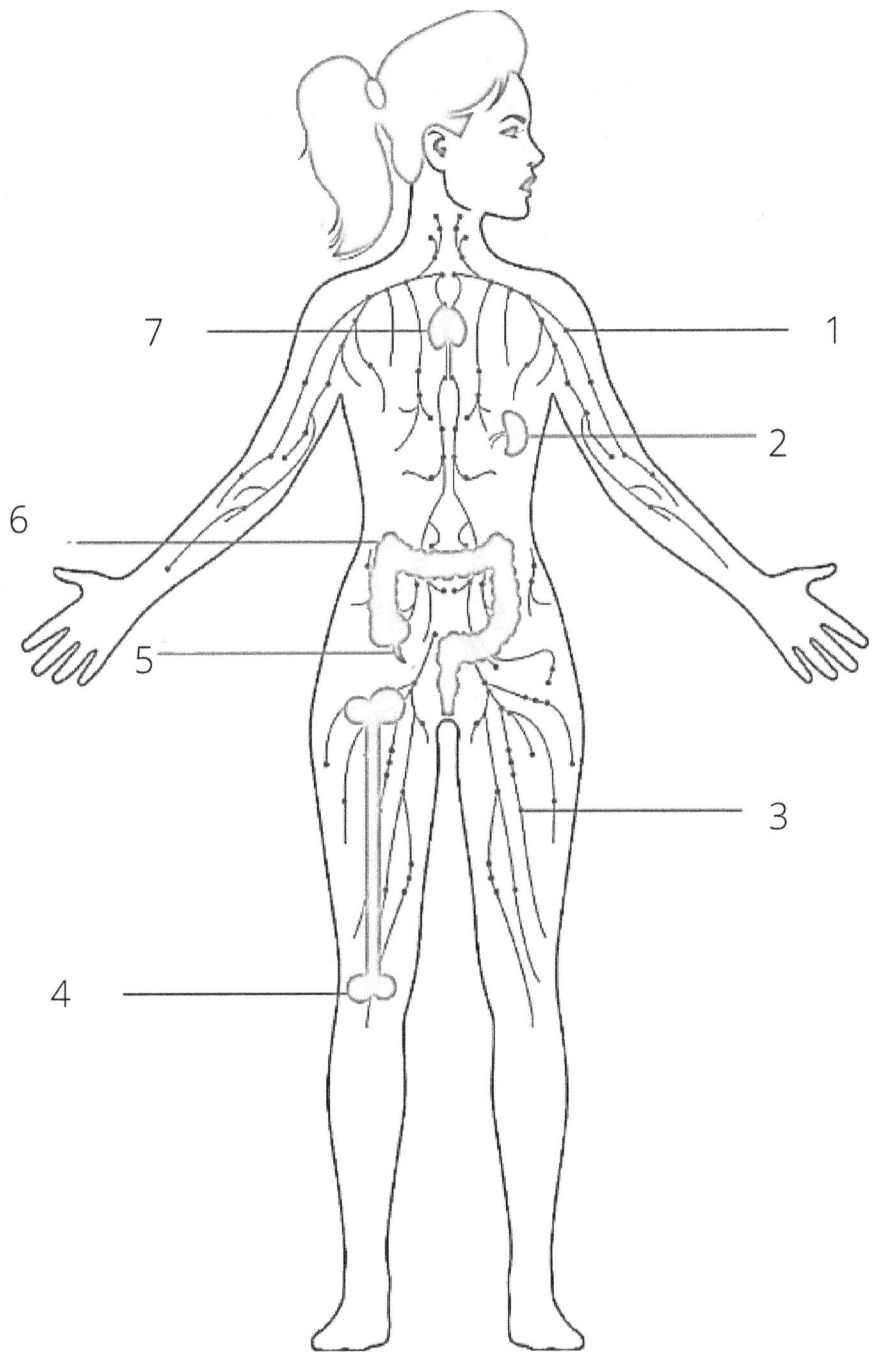

7   1

2

6

5

3

4

Lymphatic System

Select different colors for each of the areas provided with a coding numbers and corresponding structures in the diagram.

1. Cervical Lymph Nodes
2. Thymus
3. Axillary Lymph Nodes
4. Cisterna Chyli
5. Pelvic Lymph Nodes
6. Inguinal Lymph Nodes
7. Lymphatic Vessels
8. Spleen
9. Thoratic Duct
10. Tonsil

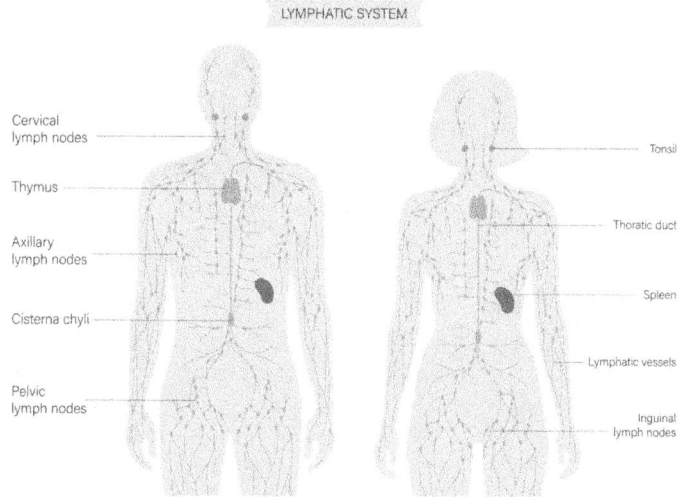

LYMPHATIC SYSTEM

Cervical lymph nodes

Thymus

Axillary lymph nodes

Cisterna chyli

Pelvic lymph nodes

Tonsil

Thoratic duct

Spleen

Lymphatic vessels

Inguinal lymph nodes

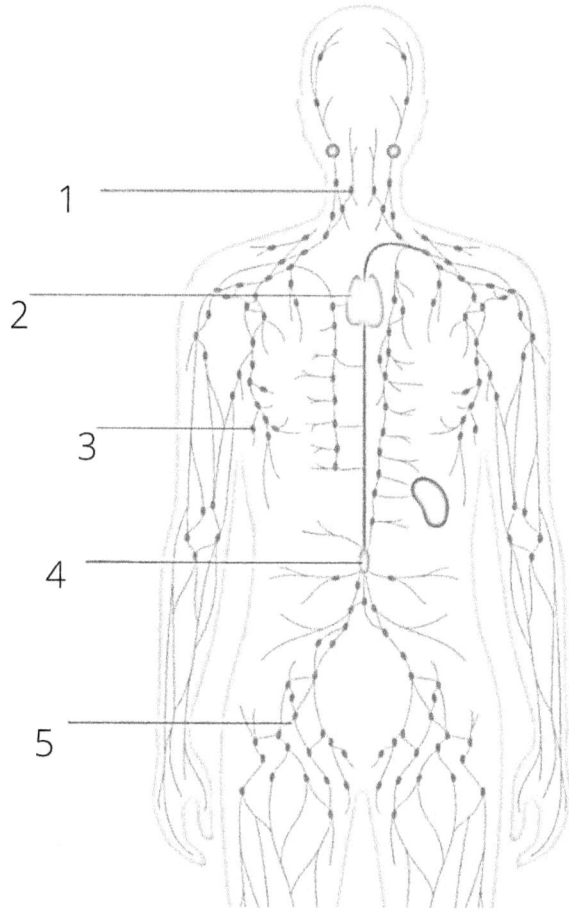
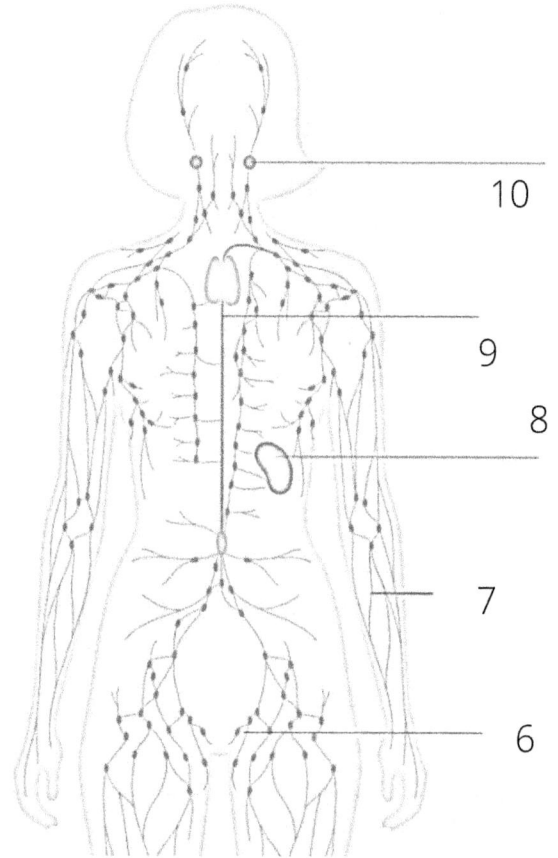

1

2

3

4

5

10

9

8

7

6

Lymph Node Anatomy

Select different colors for each of the areas provided with a coding numbers and corresponding structures in the diagram.

1. Vein
2. Artery
3. Efferent Lymphatic Vessel
4. Medulla
5. Trabeculae
6. Inner Cortex
7. Outer Cortex
8. Follicle
9. Capsule
10. Afferent Lymphatic Vessel

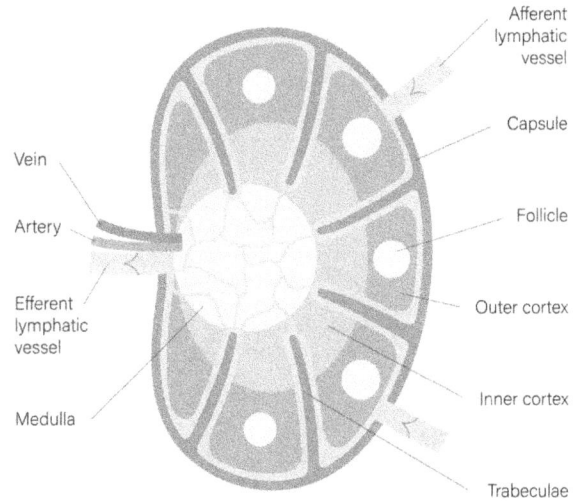

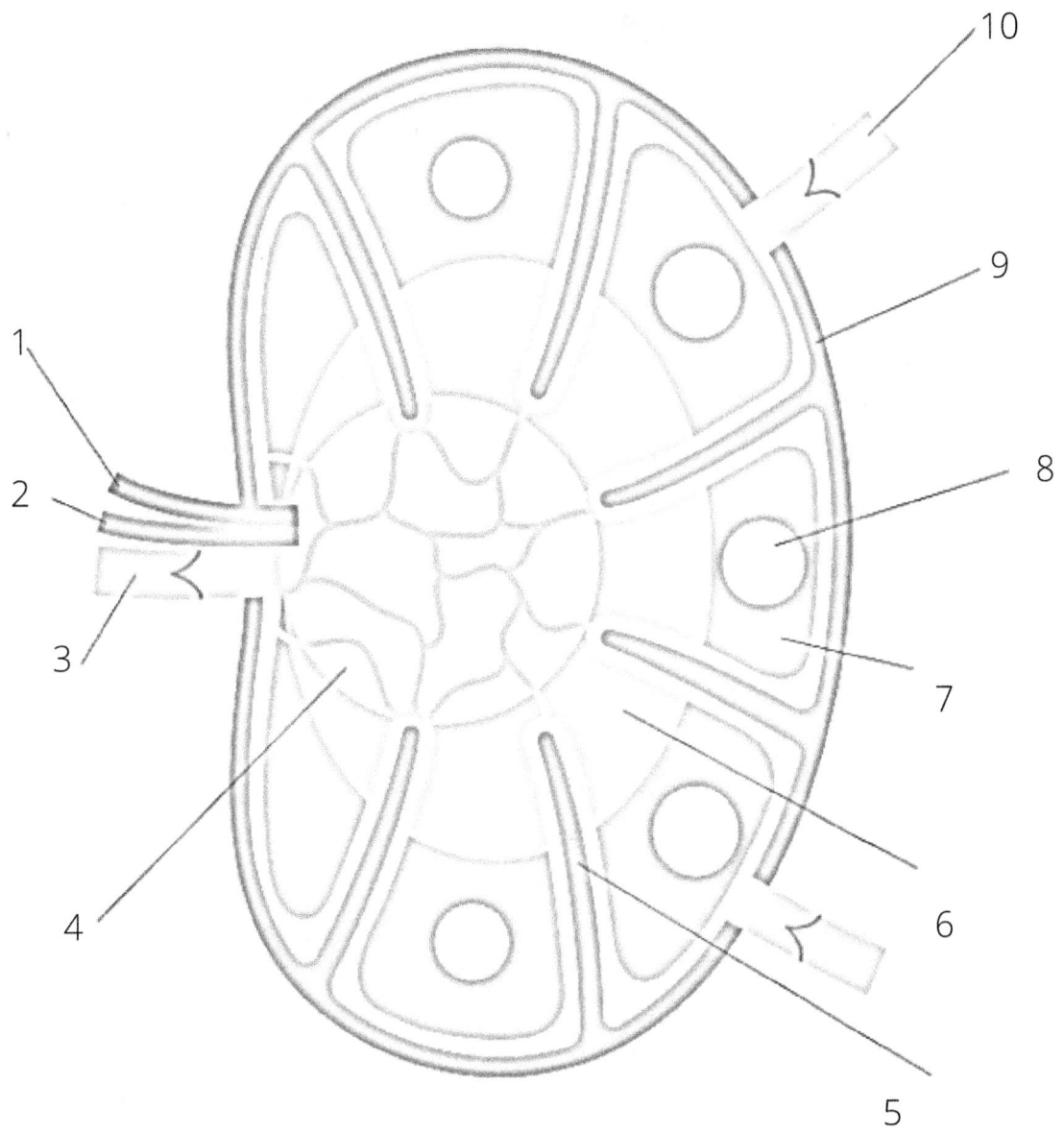

Lymph Nodes

Select different colors for each of the areas provided with a coding numbers and corresponding structures in the diagram.

1. Preauricular
2. Parotid
3. Tonsillar
4. Submental
5. Submandibular
6. Deep Cervical
7. Supraclavicular
8. Posterior Cervical
9. Superior Cervical
10. Occipital
11. Posterior Occipital

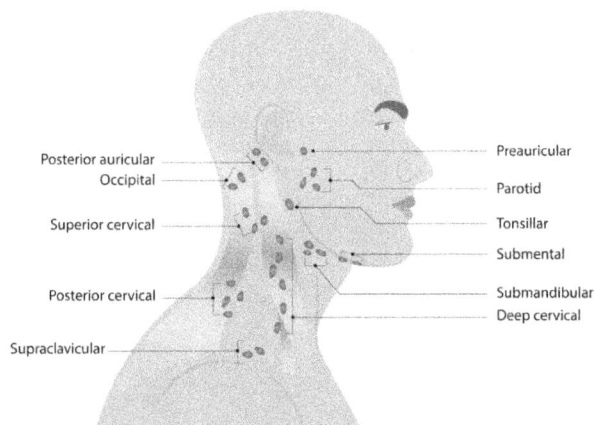

LYMPH NODES (HEAD & NECK)

Posterior auricular
Occipital
Superior cervical
Posterior cervical
Supraclavicular

Preauricular
Parotid
Tonsillar
Submental
Submandibular
Deep cervical

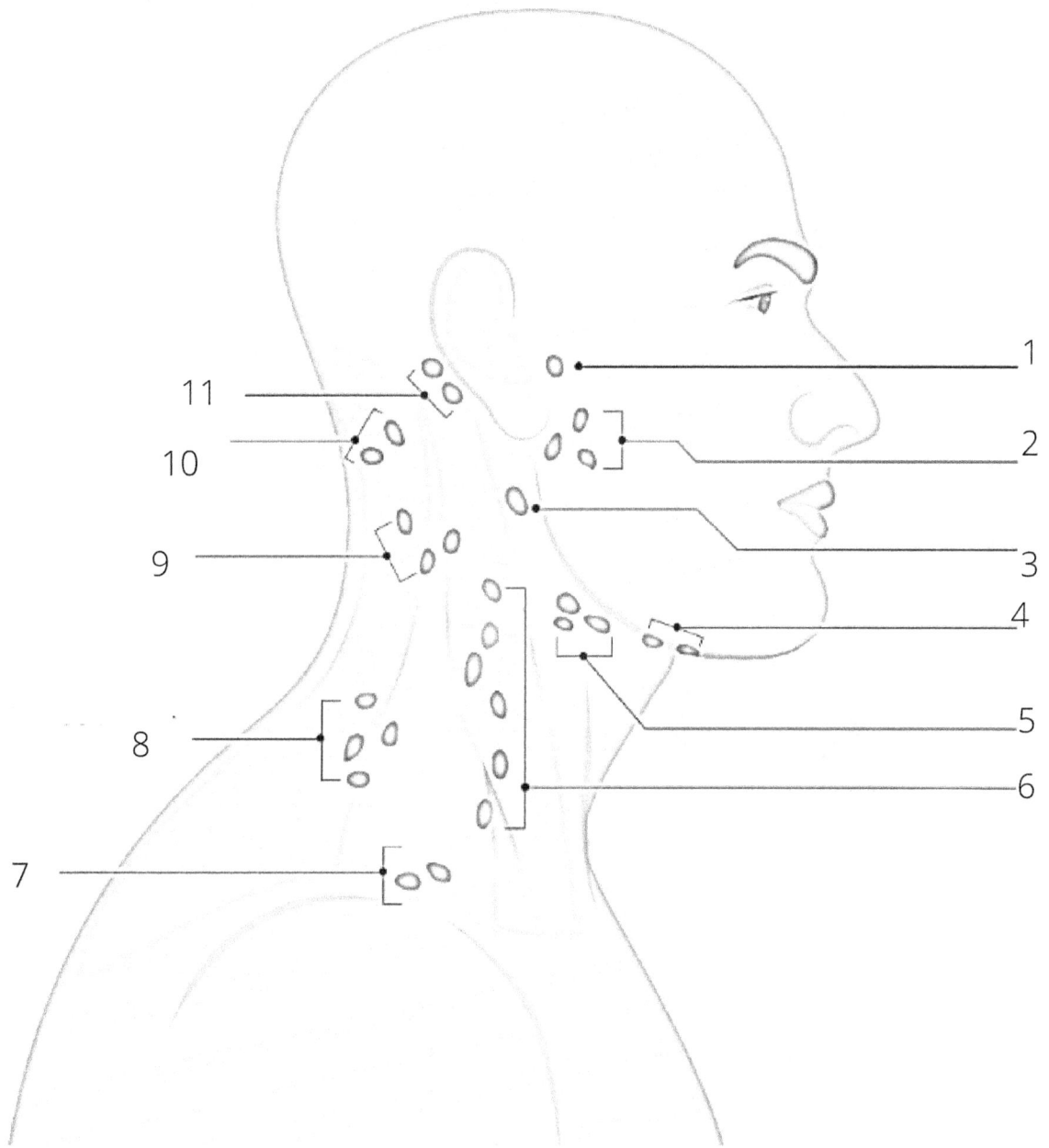

1

2

3

4

5

6

7

8

9

10

11

Lymph Node Anatomy

Select different colors for each of the areas provided with a coding numbers and corresponding structures in the diagram.

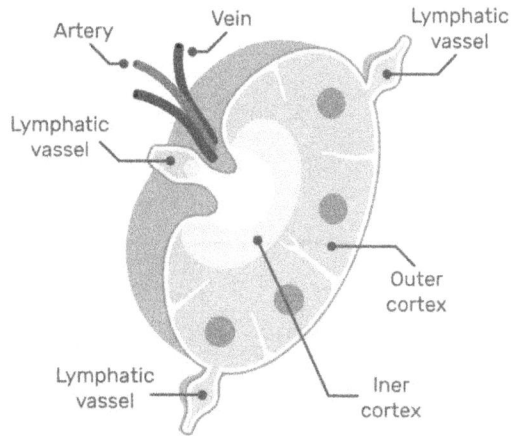

Artery

Vein

Lymphatic vassel

Lymphatic vassel

Outer cortex

Lymphatic vassel

Iner cortex

LYMPH NODE

TUMOR

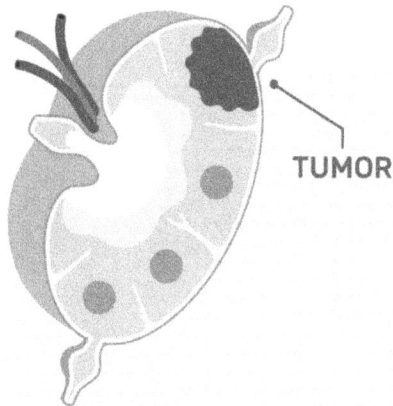

LYMPH NODE
WITH TUMOR

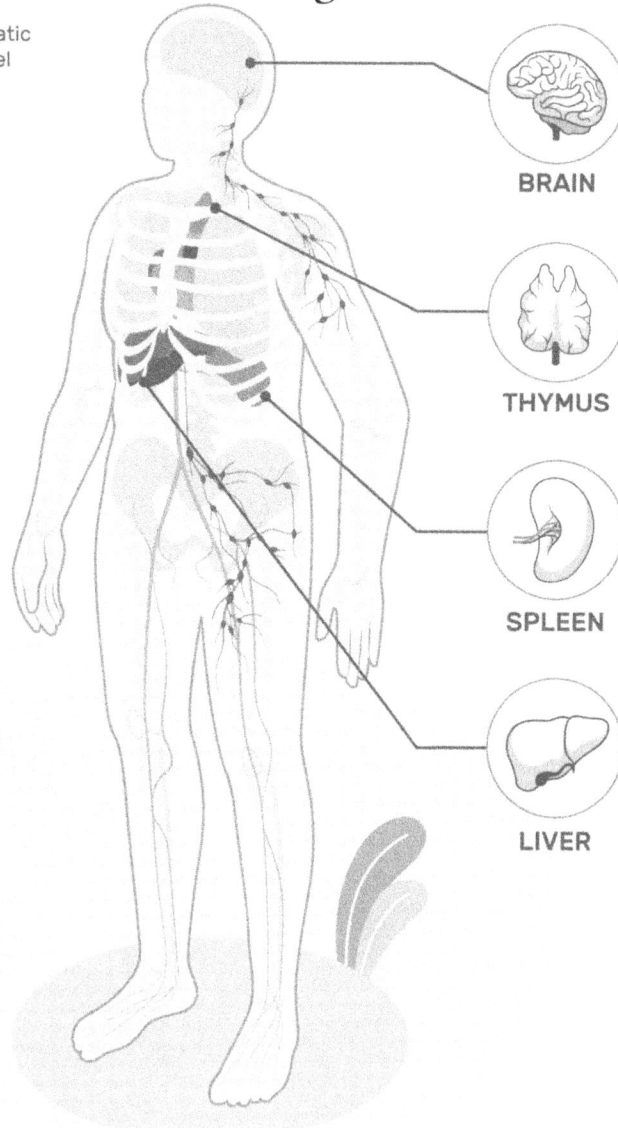

BRAIN

THYMUS

SPLEEN

LIVER

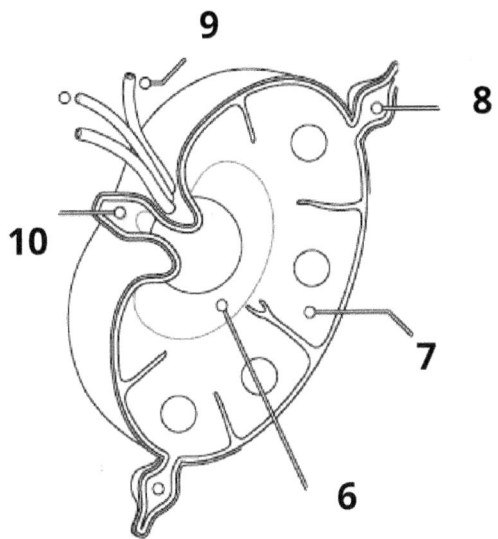

9

8

10

7

6

LYMPH NODE

5

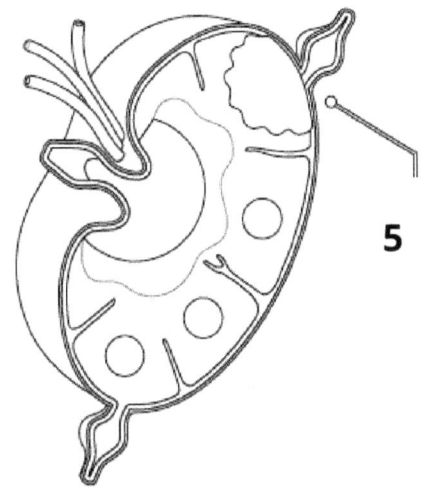

LYMPH NODE
WITH TUMOR

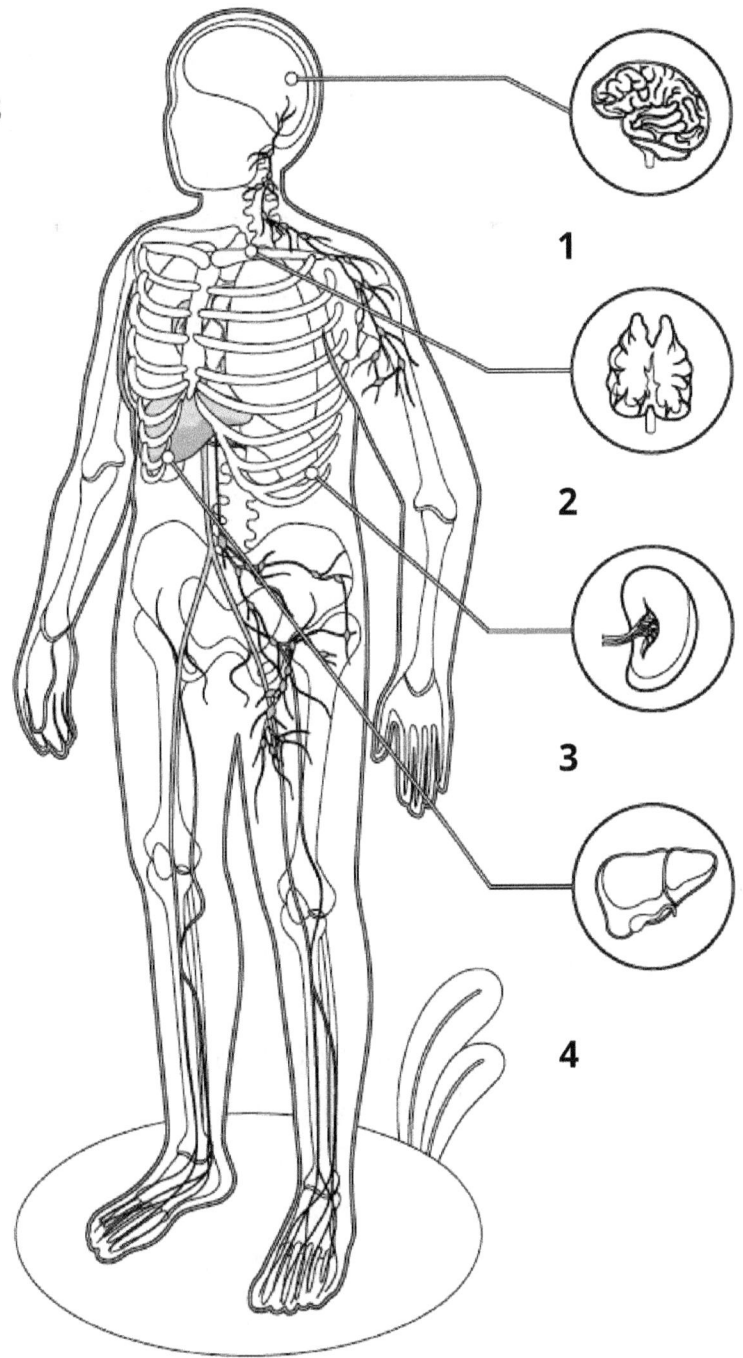

1

2

3

4

11.Respiratory System

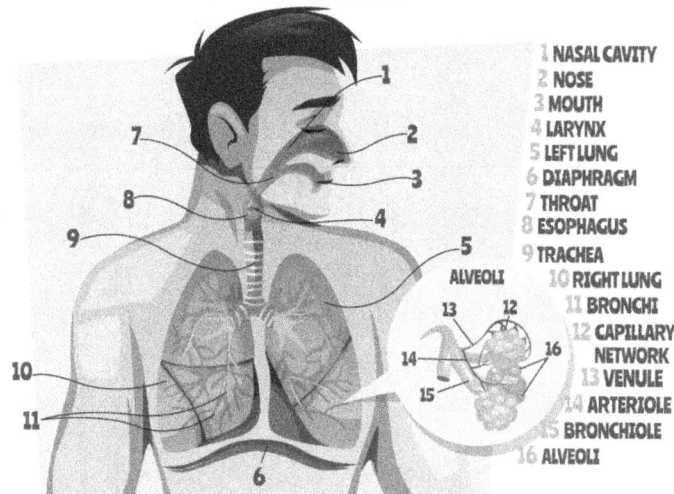

1 NASAL CAVITY
2 NOSE
3 MOUTH
4 LARYNX
5 LEFT LUNG
6 DIAPHRAGM
7 THROAT
8 ESOPHAGUS
9 TRACHEA
10 RIGHT LUNG
11 BRONCHI
12 CAPILLARY NETWORK
13 VENULE
14 ARTERIOLE
15 BRONCHIOLE
16 ALVEOLI

ALVEOLI

The Respiratory system exchanges Oxygen and CO_2 between the blood and air and comprises the lungs and passageways.

The upper respiratory tract consists of the components outside the chest cavity: the air passages of the nose, nasal cavities, throat, larynx, and top trachea.

The lower respiratory tract includes the components within the chest cavity: the lower trachea and the lungs, including the bronchial tubes and alveoli.

The main organ of the respiratory system is the lungs. Other respiratory organs include the nose, the trachea, and the breathing muscles.

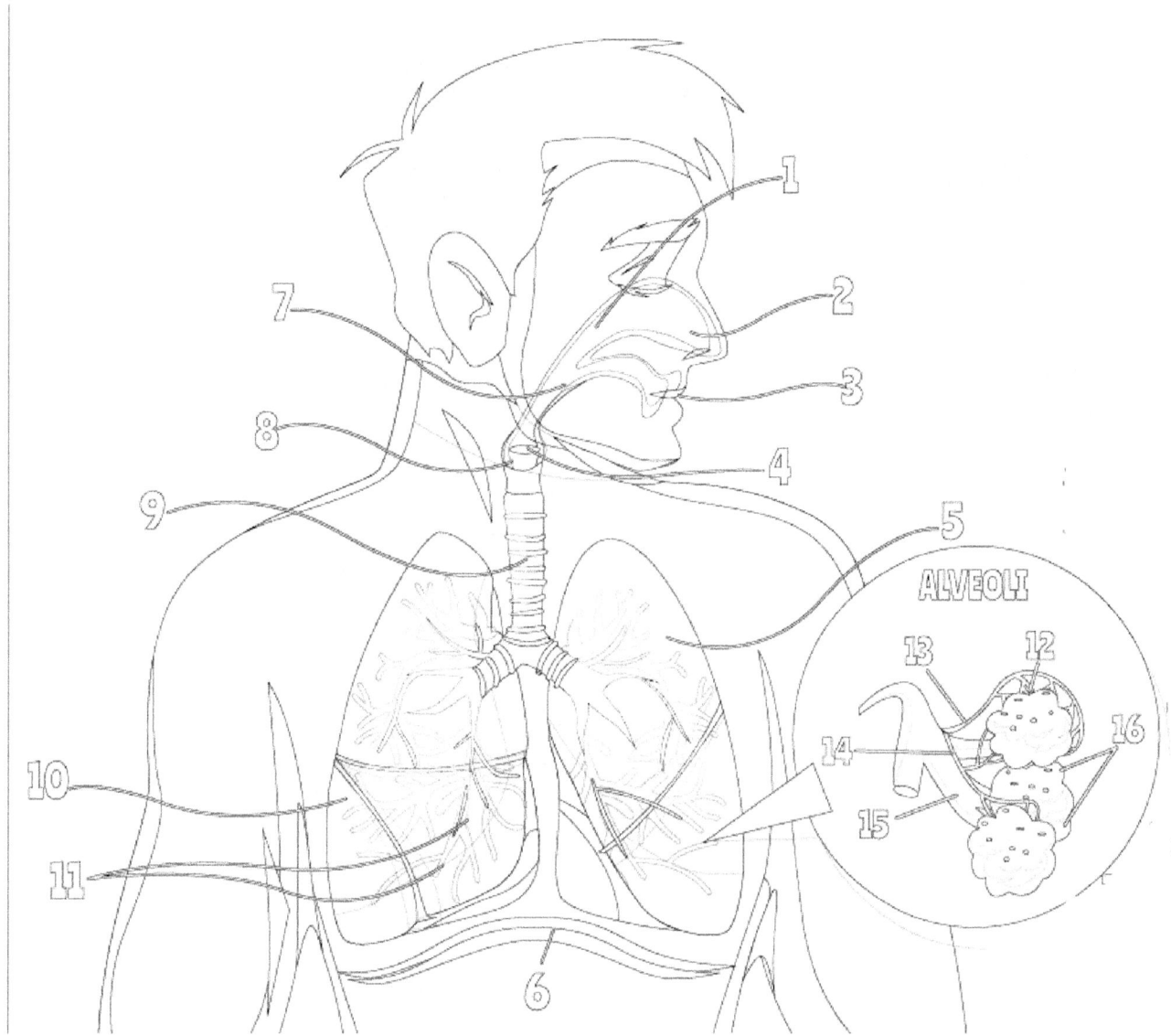

ALVEOLI

1
2
3
4
5
6
7
8
9
10
11
12
13
14
15
16

Human Larynx

Select different colors for each of the areas provided with a coding numbers and corresponding structures in the diagram.

1. Sphenoid Sinus
2. Pharingeal Tonsil
3. Eustachian Tube
4. Soft Palate   12. Vestibule
5. Oral Cavity   13. Inferior Nasal Concha
6. Epiglottis   14. Middle Nasal Concha
7. Hyoid Bone 15. Superior Nasal Concha
8. Mandible   16. Olfactory Nerves
9. Tongue   17. Olfactory Buld
10. Lips   18. Frontal Sinus
11. Hard Palate

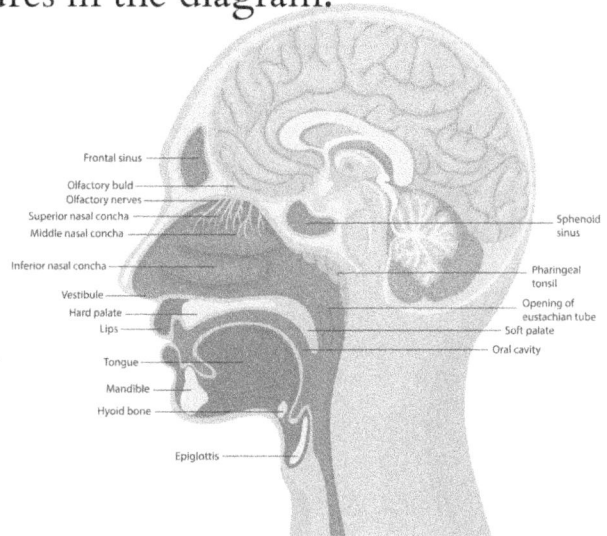

Frontal sinus
Olfactory buld
Olfactory nerves
Superior nasal concha
Middle nasal concha
Inferior nasal concha
Vestibule
Hard palate
Lips
Tongue
Mandible
Hyoid bone
Epiglottis
Sphenoid sinus
Pharingeal tonsil
Opening of eustachian tube
Soft palate
Oral cavity

_____

_____

_____

_____

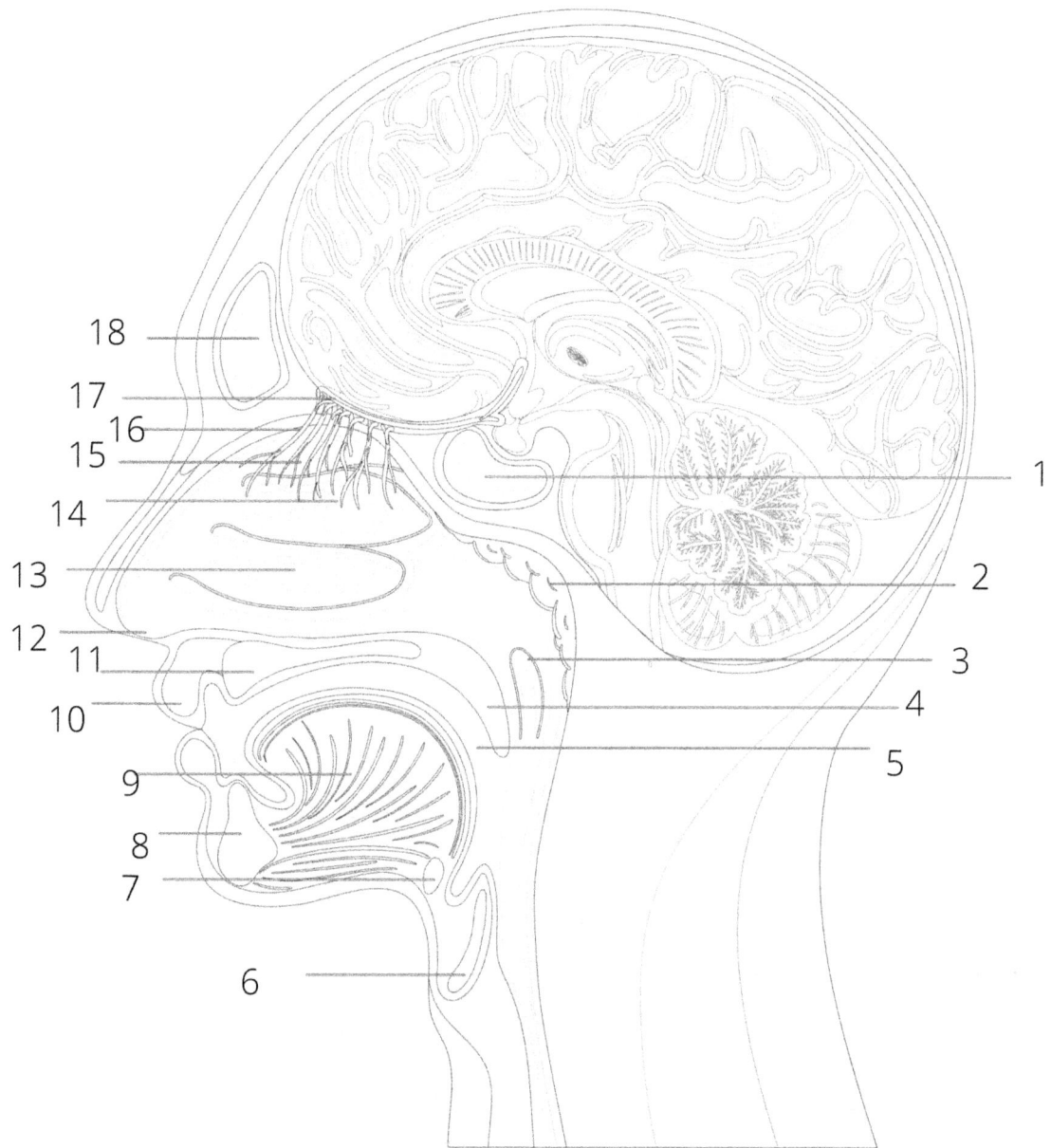

Lung

Select different colors for each of the areas provided with a coding numbers and corresponding structures in the diagram.

1. Pleura
2. Primary Bronchi
3. Right Lung
4. Tertiary Bronchi
5. Left Lung
6. Secondary Bronchi
7. Trachea

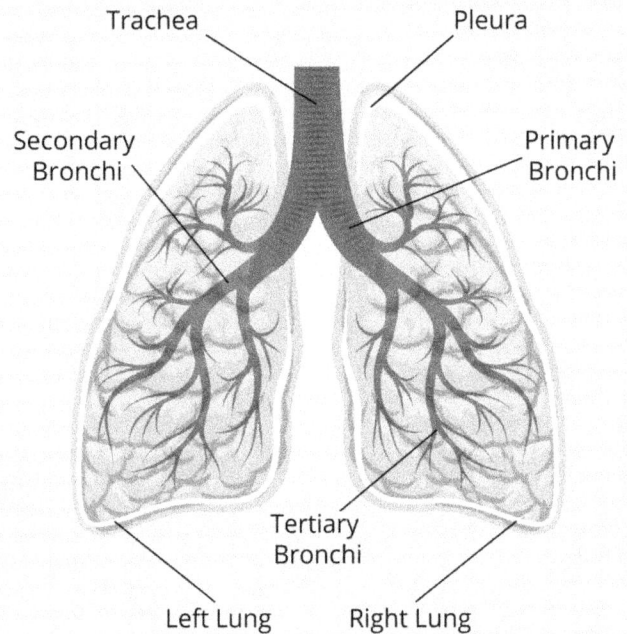

Trachea

Pleura

Secondary Bronchi

Primary Bronchi

Tertiary Bronchi

Left Lung

Right Lung

_____

_____

_____

_____

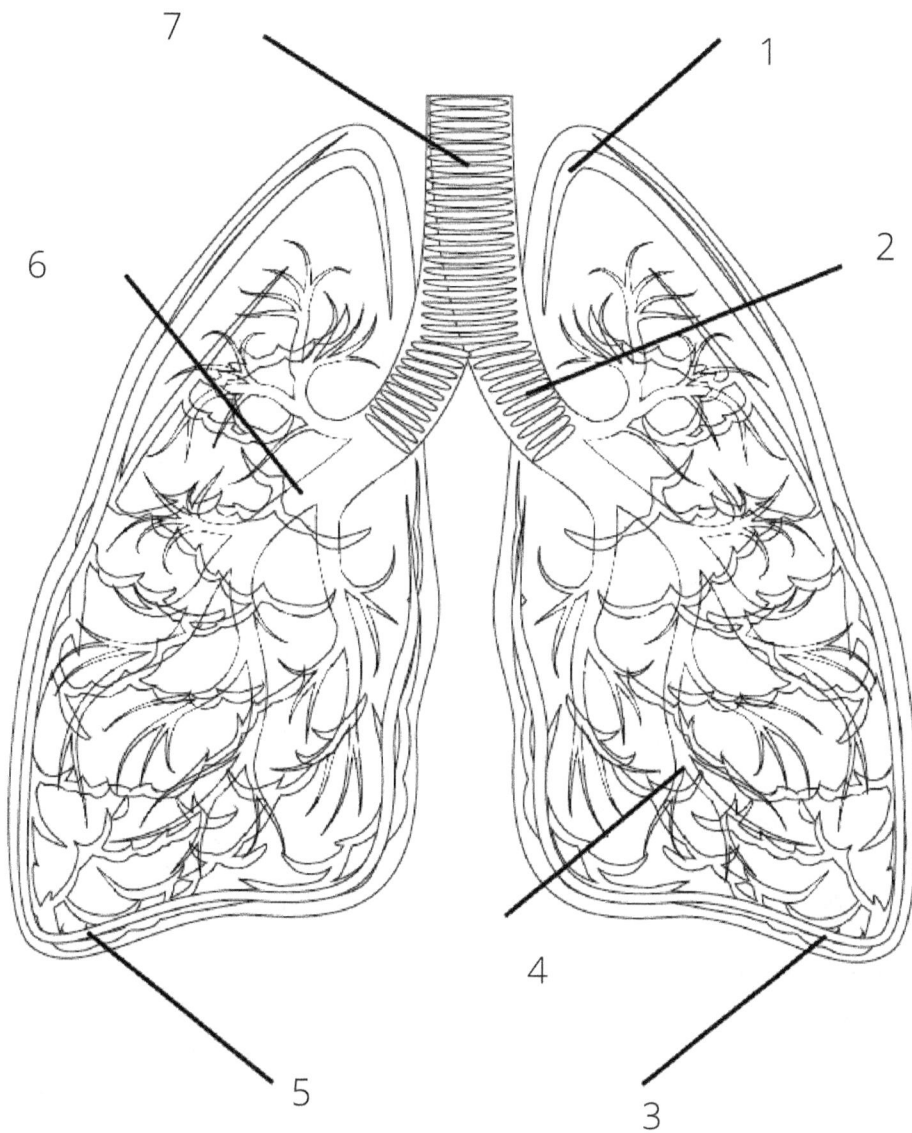

Breathing

Select different colors for each of the areas provided with a coding numbers and corresponding structures in the diagram.

1. Inhalation
2. Exhalation

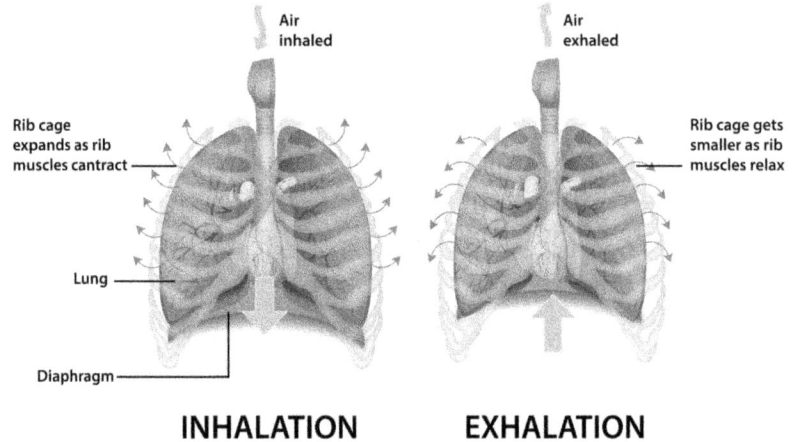

Air inhaled

Rib cage expands as rib muscles cantract

Lung

Diaphragm

INHALATION

Air exhaled

Rib cage gets smaller as rib muscles relax

EXHALATION

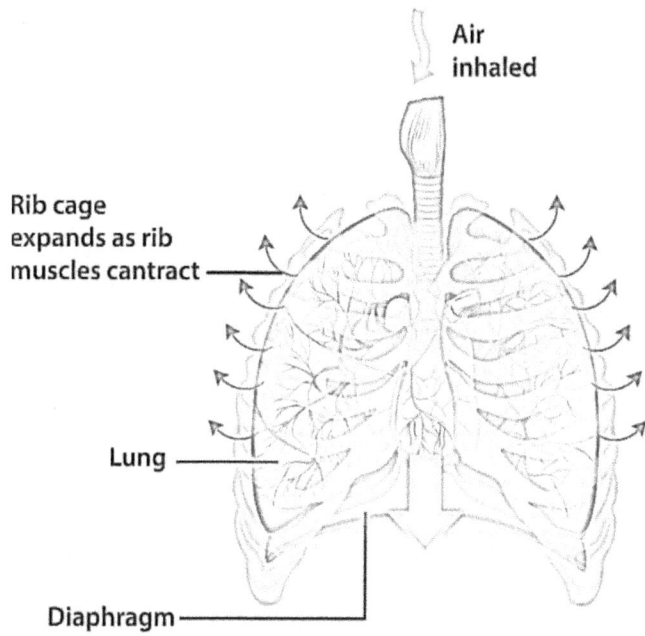

Air
inhaled

Rib cage
expands as rib
muscles cantract

Lung

Diaphragm

1 _____

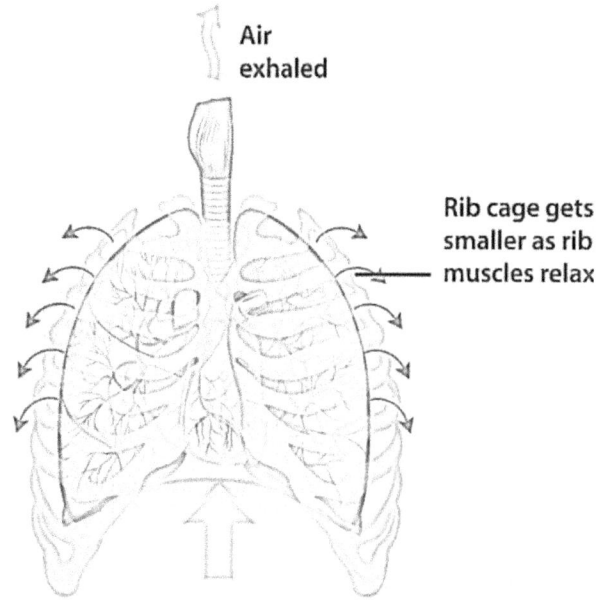

Air
exhaled

Rib cage gets
smaller as rib
muscles relax

2 _____

12.Digestive System

1. Liver
2. Gallbladder
3. Duodenum
4. Right Colic Flexure
5. Duodeno Jejunal Junction
6. Ascending Colon
7. Ileocecal Junction
8. Ileum
9. Cecum
10. Appendix
11. Stomach
12. Pancreas
13. Left Colic Flexure
14. Transverse Colon
15. Jejunum
16. Descending Colon
17. Sigmoid Colon
18. Rectum
19. Anal Canal

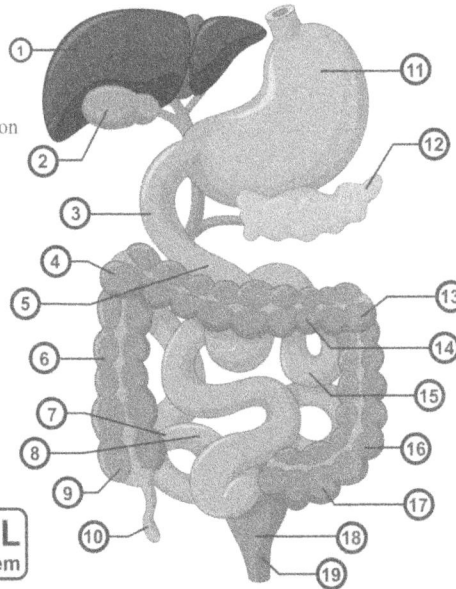

INTERNAL
Human Digestive System

The organs take in food and liquids and break them down into substances that the body can use for energy, growth, and tissue repair. Waste products the body cannot use leave the body through bowel movements.

The part of the Digestive System includes the mouth, pharynx (throat), esophagus, stomach, small intestine, large intestine, rectum, and anus.

Major Functions: Motility, digestion, absorption, and secretion

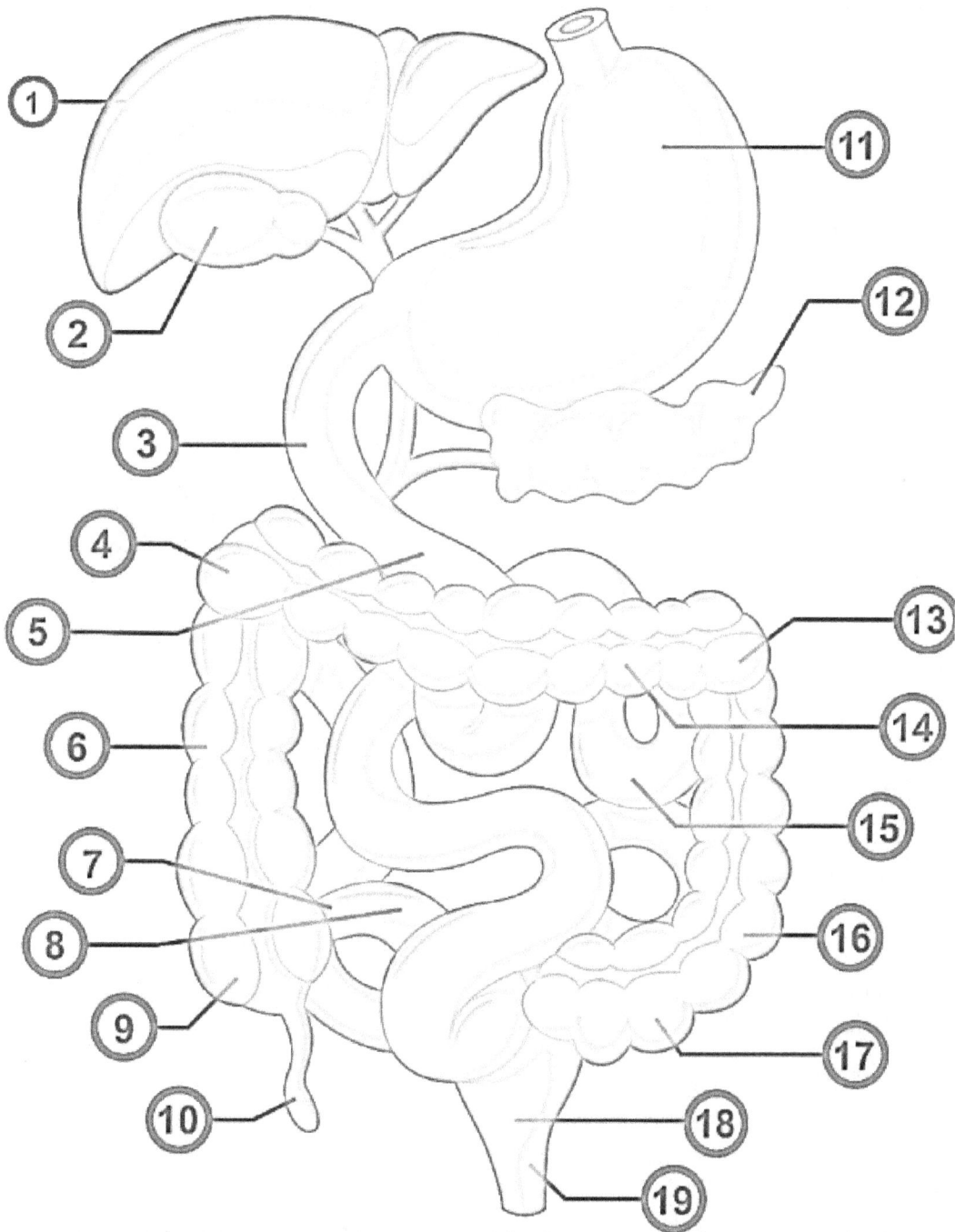

Human Digestive System

Select different colors for each of the areas provided with a coding numbers and corresponding structures in the diagram.

1. Salivary Glands
2. Pharynx
3. Esophagus
4. Stomach
5. Pancreas
6. Small Intestine
7. Rectum
8. Anus
9. Appendix
10. Large Intestine
11. Gal Bladder
12. Liver
13. Epiglottis
14. Tongue
15. Mouth

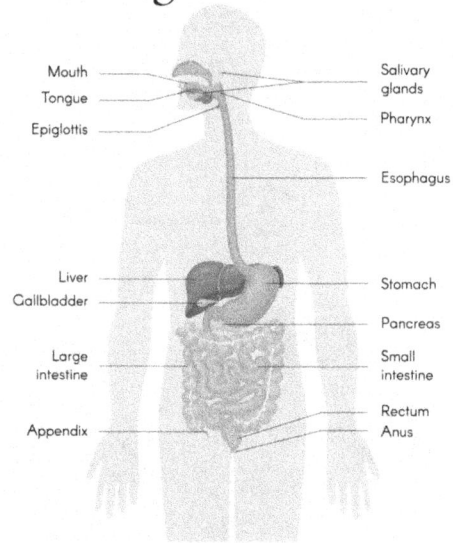

Mouth
Tongue
Epiglottis
Liver
Gallbladder
Large intestine
Appendix
Salivary glands
Pharynx
Esophagus
Stomach
Pancreas
Small intestine
Rectum
Anus

_____

_____

_____

_____

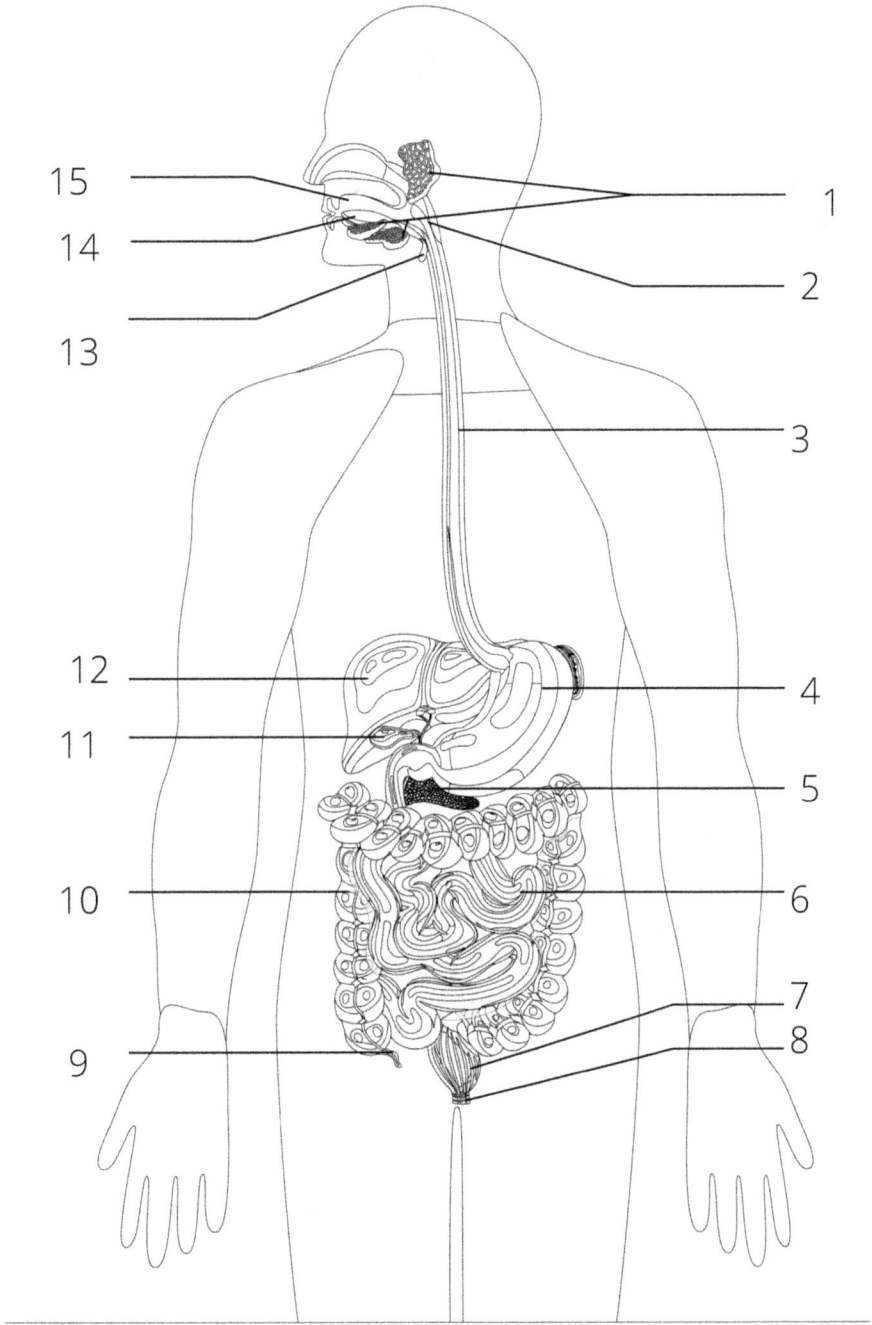

Digestion Process

Select different colors for each of the areas provided with a coding numbers and corresponding structures in the diagram.

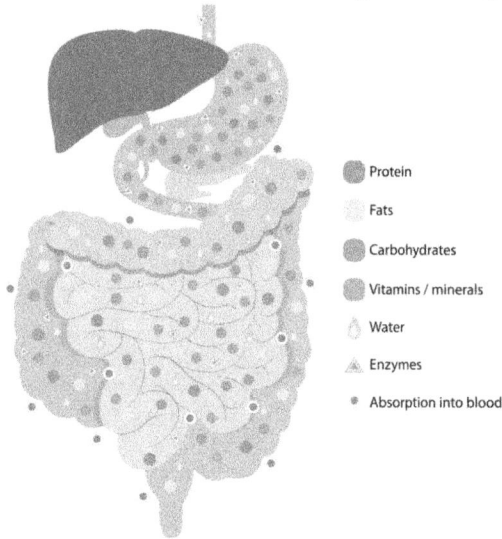

- Protein
- Fats
- Carbohydrates
- Vitamins / minerals
- Water
- Enzymes
- Absorption into blood

1. Protein
2. Fats
3. Carbohydrates
4. Vitamins/Minerals
5. Water
6. Enzymes
7. Absorption into blood

_____

_____

_____

_____

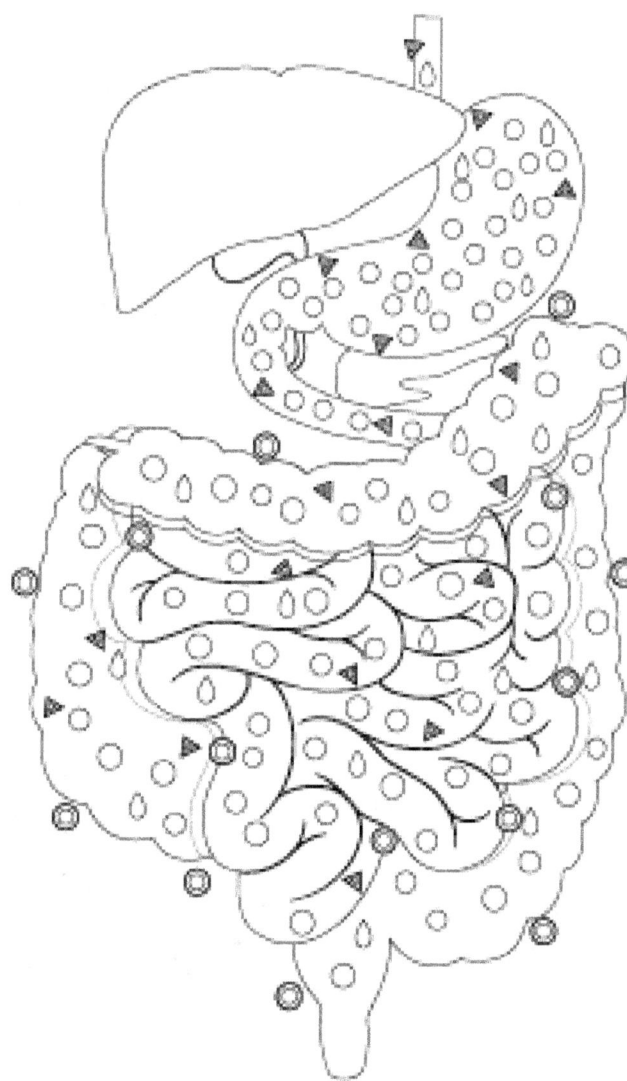

1
2
3
4
5
6
7

13.Urinary System

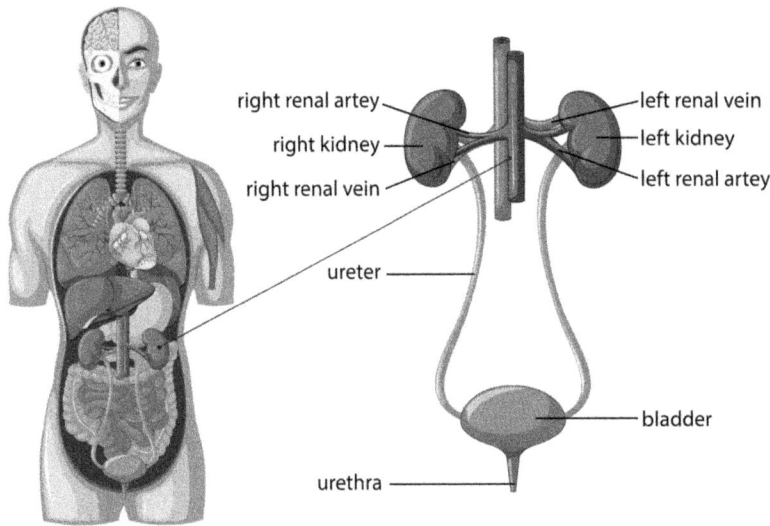

right renal artey
right kidney
right renal vein

left renal vein
left kidney
left renal artey

ureter

bladder

urethra

1. Left Renal Vein
2. Left Kidney
3. Left Renal Artery
4. Bladder
5. Urethra
6. Ureter
7. Right Renal Vein
8. Right Kidney
9. Right Renal Artery

The urinary system consists of two kidneys, two ureters, the urinary system bladder, and the urethra.

The urine formation is the feature of the kidneys, and the rest of the system is in charge of removing the urine.

It gets rid of wastes from the blood and also aids in maintaining water and electrolyte balance.

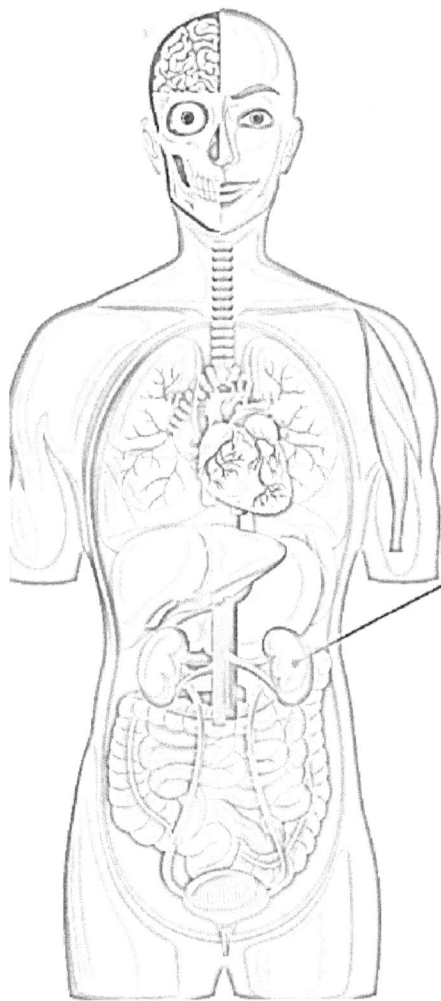
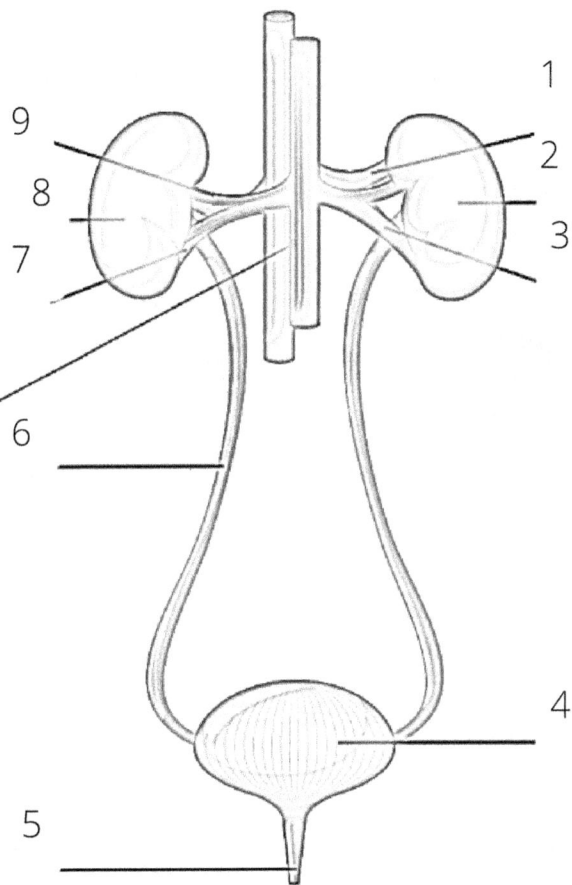

9

8

7

1

2

3

6

4

5

Kidney Anatomy

Select different colors for each of the areas provided with a coding numbers and corresponding structures in the diagram.

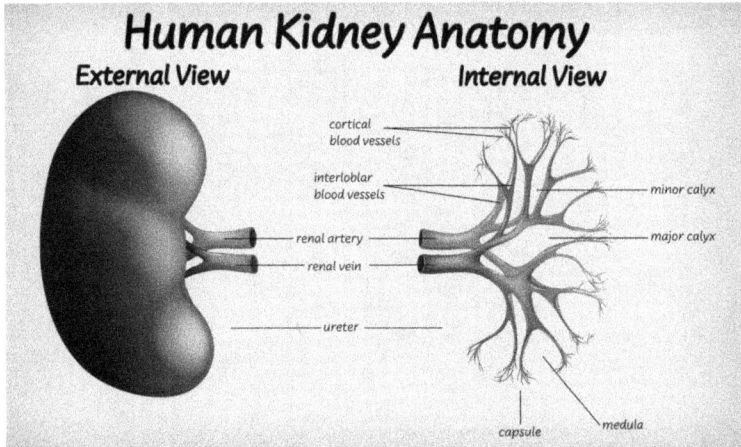

Human Kidney Anatomy
External View   Internal View
cortical blood vessels
interloblar blood vessels
renal artery
renal vein
ureter
minor calyx
major calyx
capsule   medula

1. Minor Calyx
2. Major Calyx
3. Medula
4. Capsule
5. Ureter
6. Renal Vein
7. Renal Artery
8. Interlobar Blood Vessels
9. Cortical Blood Vessels

_____

_____

_____

_____

External View

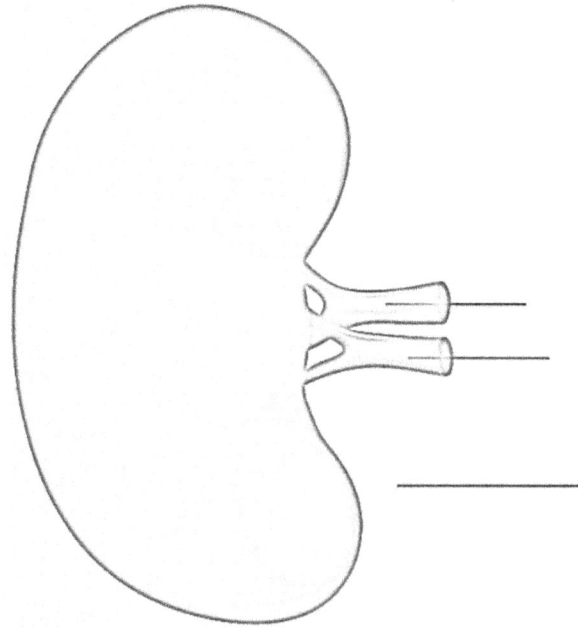

Internal View

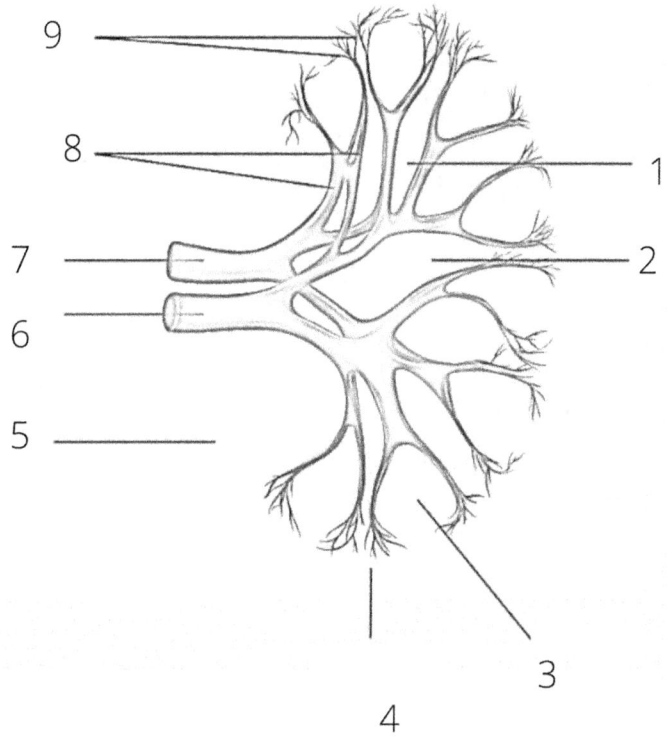

9

8

7

6

5

1

2

3

4

14. Reproductive System

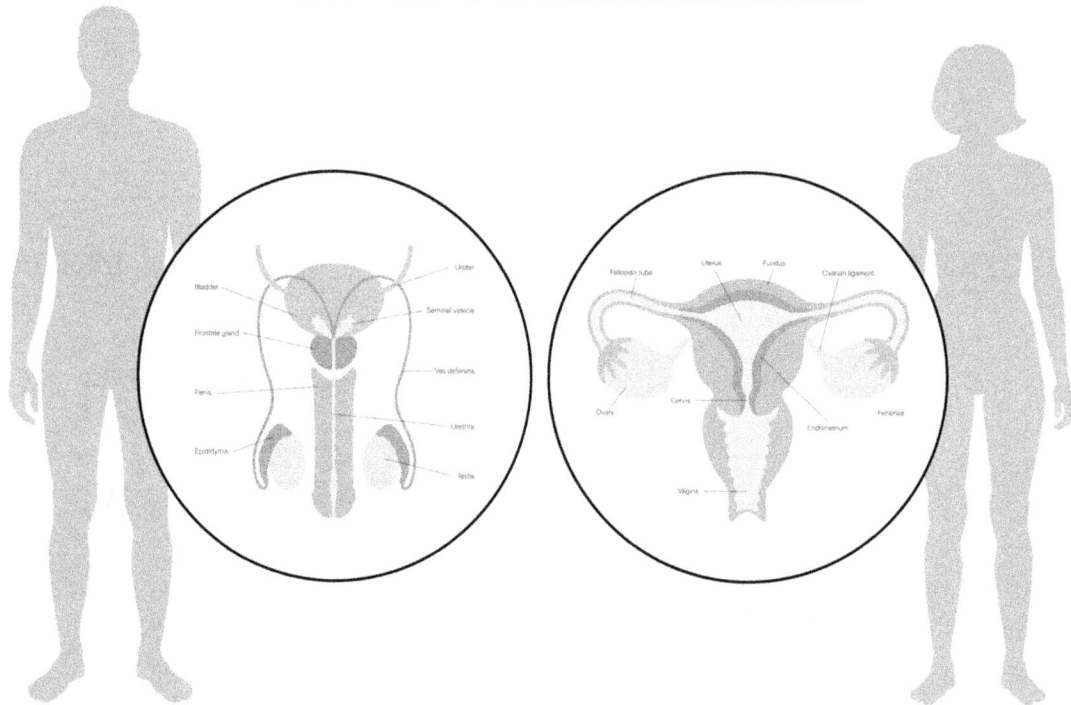

The Reproductive system provides continuity to civilization by producing offspring.

The male reproductive system contains the testes, penis, accessory organs, and vessels that produce and conduct sperm to the female reproductive tract.

The woman's reproductive system includes ovaries, uterine tubes, uterus, vagina, as well as external genitalia. She generates egg cells and also houses the infant.

Male Reproductive System

Select different colors for each of the areas provided with a coding numbers and corresponding structures in the diagram.

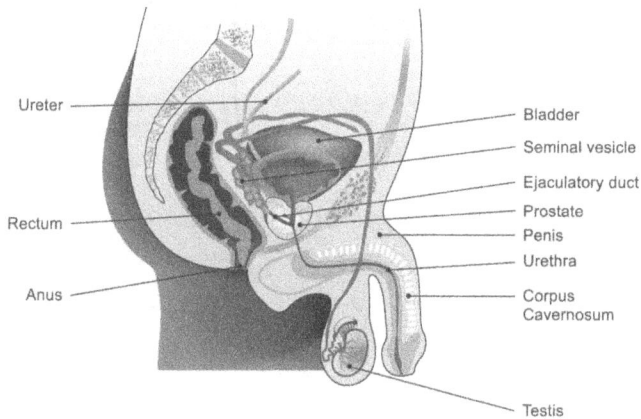

Ureter

Bladder

Seminal vesicle

Ejaculatory duct

Prostate

Rectum

Penis

Urethra

Anus

Corpus
Cavernosum

Testis

1. Bladder
2. Seminal Vesicle
3. Ejaculatory Duct
4. Prostrate
5. Penis
6. Urethra
7. Corpus Cavernosum
8. Testis
9. Anus
10. Rectum
11. Ureter

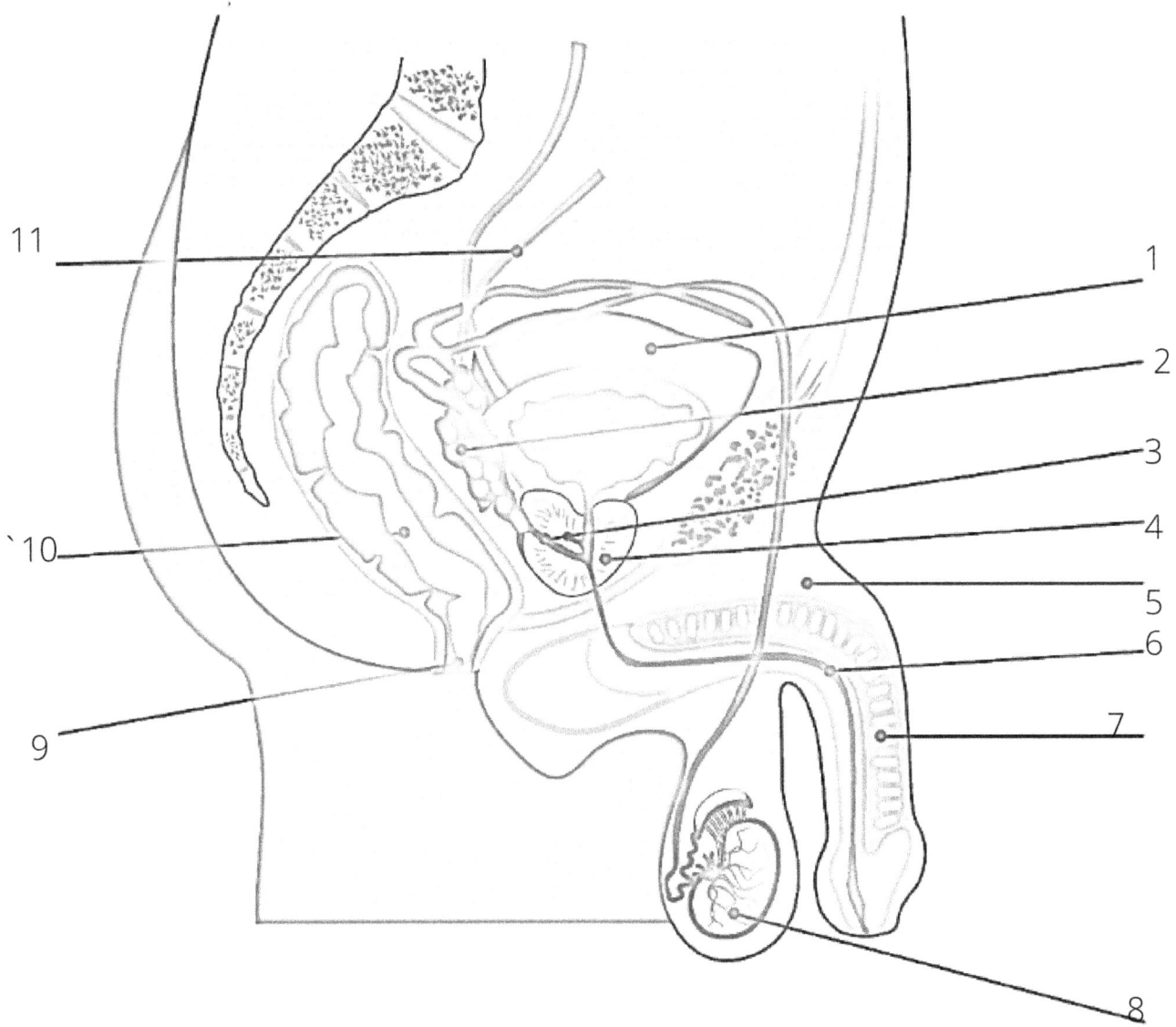

11

1

2

3

4

5

6

7

`10

9

8

Female Reproductive System

Select different colors for each of the areas provided with a coding numbers and corresponding structures in the diagram.

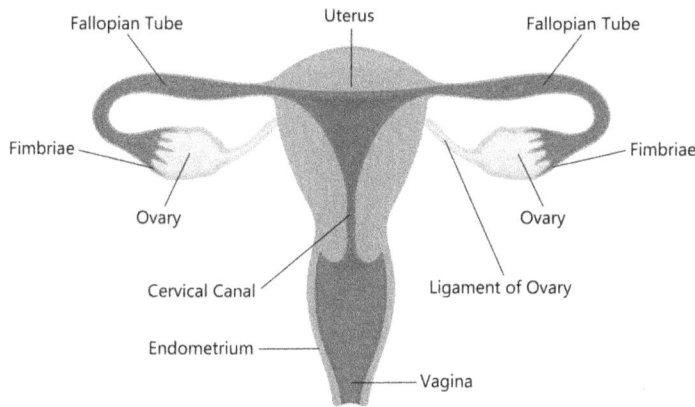

1. Uterus
2. Fallopian Tube
3. Fimbriae
4. Ovary
5. Ligament of Ovary
6. Vagina
7. Endometrium
8. Cervical Canal
9. Ovary
10. Fimbriae
11. Fallopian Tube

_____

_____

_____

_____

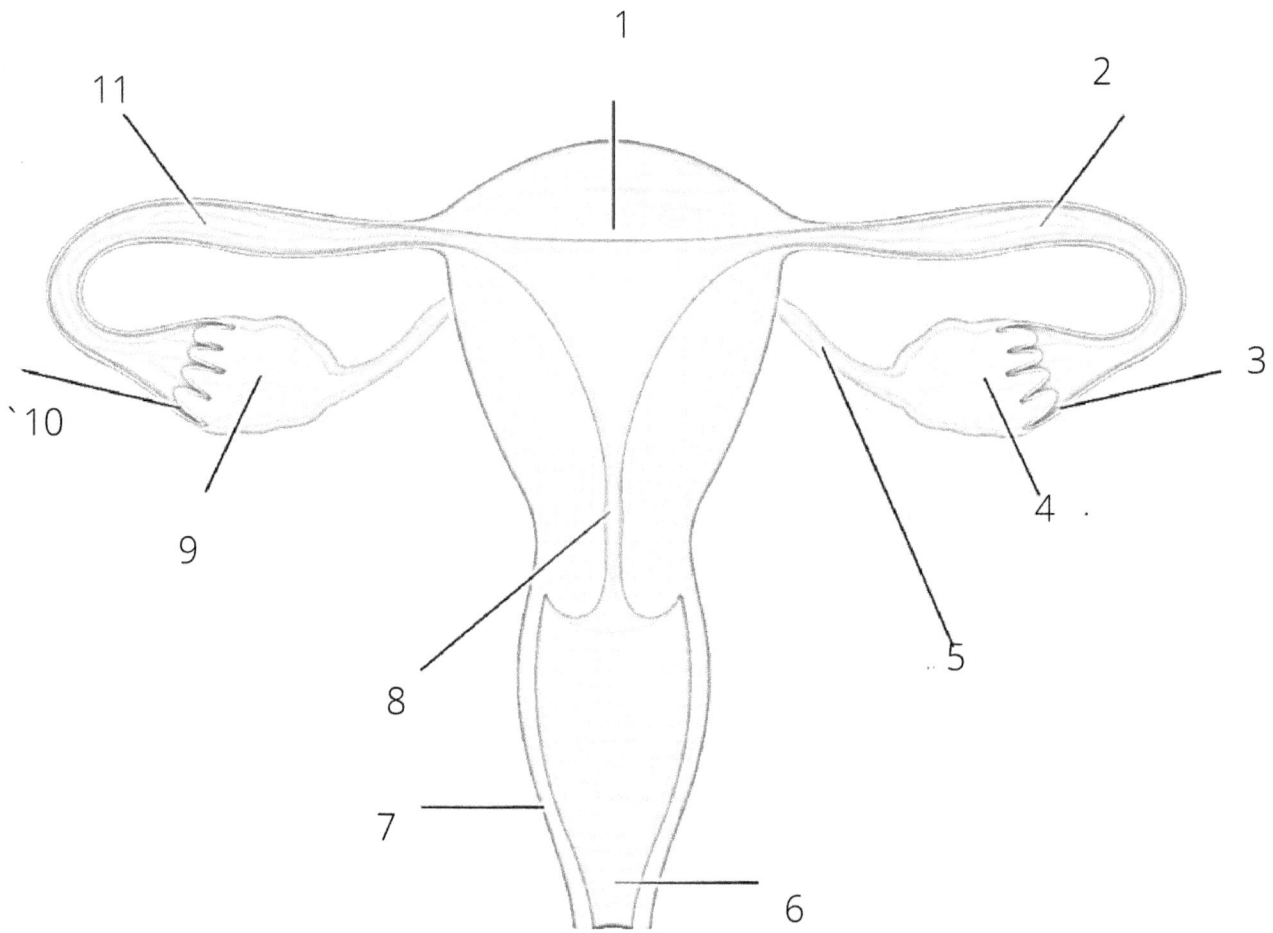

Ovary Anatomy

Select different colors for each of the areas provided with a coding numbers and corresponding structures in the diagram.

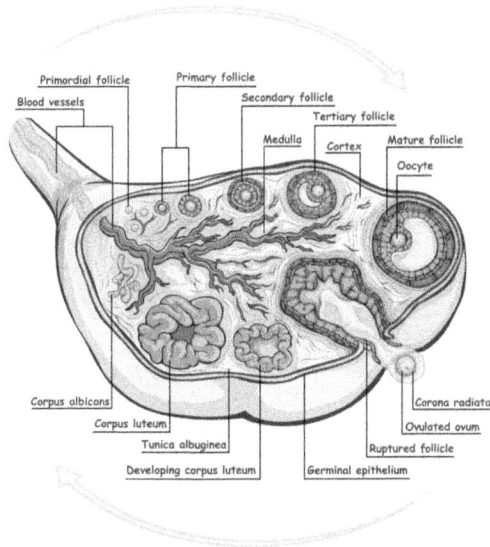

1. Blood Vessels
2. Primordial Follicle
3. Primary Follicle
4. Secondary Follicle
5. Medulla
6. Tertiary Follicle
7. Cortex
8. Mature Follicle
9. Oocyte
10. Corona Radiata
11. Ovulated Ovum
12. Ruptured Follicle
13. Germinal Epithelium
14. Developing Corpus Luteum
15. Tunica Albuginea
16. Corpus Luteum
17. Corpus Albicans

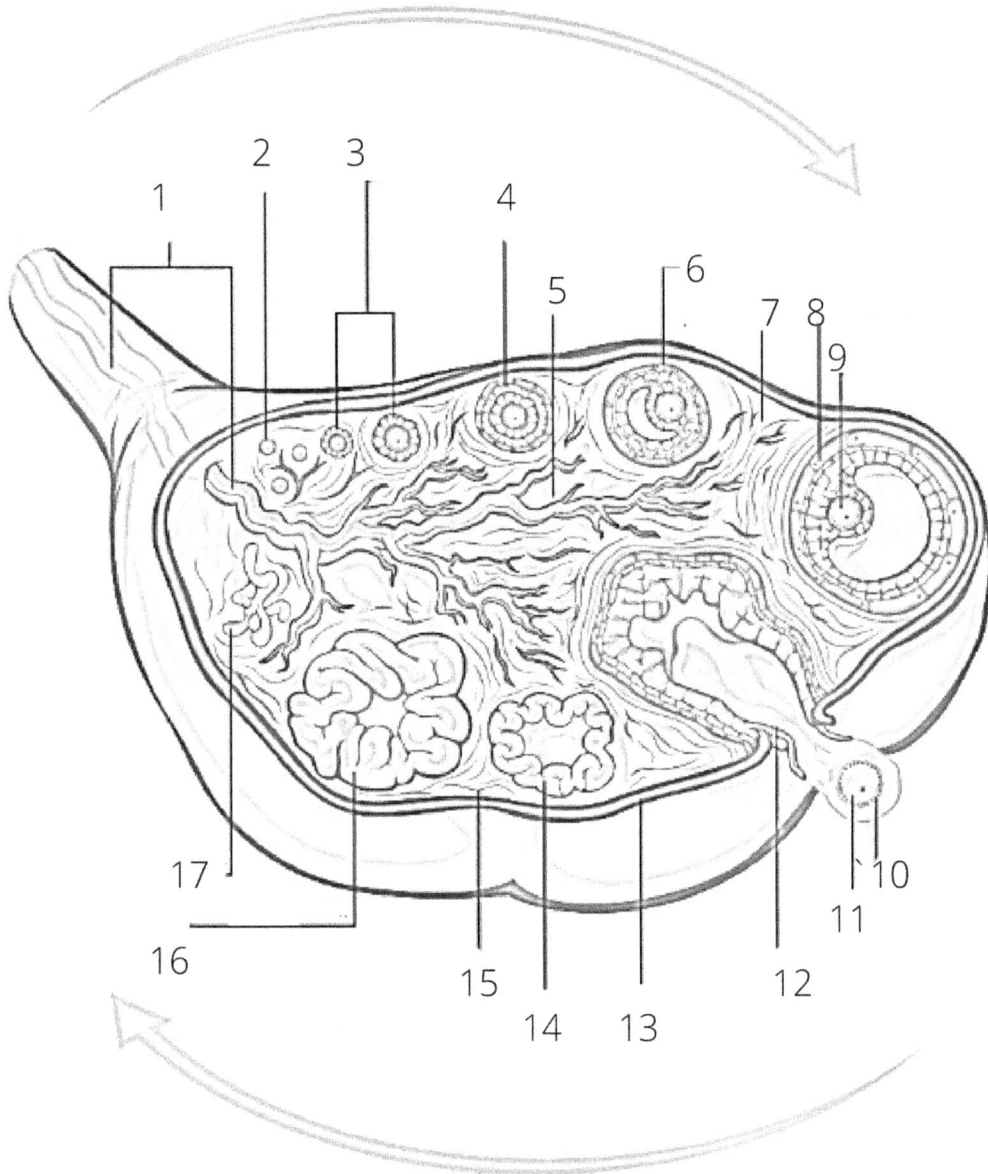

1) What is the number of bones an adult human skeleton have?

a) 135
b)206
c) 201
d) 89

2) Name the organ where the digestion of proteins takes place?

a) Pancreas
b) Liver
c)Lung
d) IIeum

MULTIPLE CHOICE QUESTIONS (MCQS)

3) Which part of the skull is called the helmet of the brain?

a) Thalamus

b) Medulla

c) Cranium

d) Pineal

4) Which of these keep blood flowing unidirectionally in humans?

a) Valves

b) Neuron

c) Veins

d) Artery

5) Which of them saves lungs?

a) Backbone
b) Ribcage
c) Sternum
d) All of them

6) Which is the functional unit of the human nervous system??

a) Neuron
b) Dendron
c) Metre
d) All of these

Multiple Choice Questions (MCQs)

7) What are the types of bone marrow?

a) Red and white marrow
b) Yellow Marrow
c) Both A & B
d) Red marrow and yellow marrow

8) What are the contents of the ventral cavity?

a) Heart
b) Lung
c) Spinal Cord
d) viscera

9) What does a connective tissue does?

   a) Form a packing around the organs
   b) Binding tissues
   c) Support parts of the human body
   d) All of these

10) The study of Tissues is known as?

a) Anatomy
b) Physiology
c) Pathology
d) Histology

MULTIPLE CHOICE QUESTIONS (MCQs)

11) Which part of the brain controls the body temperature and appetite?

a) Medulla
b) Cerebellum
c) Hypothalamus
d) Neuron

12) What are Osteocytes?

a) Gland
b) Tissue
c) Bone cells
d) None of these

MULTIPLE CHOICE QUESTIONS (MCQS)

13)Fine movement of the hand is caused by which nerve?

a) Radial
b) Median
c) Femur
d) Ulnar

14)Study of blood vessels is known as?

a) Angiology
b) Radiology
c)Pulmonology
d) Cardiology

MULTIPLE CHOICE QUESTIONS (MCQS)

15) What is hemoglobin?

a) Protein
b) Respiratory Pigment
c) Nerve
d) Tissue

16) The main artery Aorta originates from?

a) Auricle
b) Right Ventricle
c) Left Ventricle
d) None of these

MULTIPLE CHOICE QUESTIONS (MCQS)

17) What are the four types of bones?

a) Conical,Flat,Small,Large
b) Regular, Irregular, Flat, Inclined
c)Big ,Small,Flat,Regular
d)Long, Short, Flat, Irregular

18) Which one of these covers the crown of teeth?

a) Enamel
b) Root Canal
c) Nucleosides
d) Polysaccharides

MULTIPLE CHOICE QUESTIONS (MCQS)

19) Which among the below options is a breast bone?

a) Rib
b) Sternum
c) Cranium
d) Scapula

20) How many bones are there in a Human Cranium?

a) 5
b) 7
c) 6
d) 8

Multiple Choice Questions (MCQs)

21) How many bones are there in a Human Skull?
a) 31
b) 29
c) 44
d) 34

22) Narrow tubes in the lungs are called ....?

a) Bronchi
b) Bronchioles
c) Veins
d) Trachea

MULTIPLE CHOICE QUESTIONS (MCQS)

23) This helps human sperm to move _____?

a) Nucleoid
b) Basal body
c) Cilia
d) Flagellum

24) What passes through the Bidder's canal?

a) Blood
b) Proteins
c) Sperm
d) None of these

MULTIPLE CHOICE QUESTIONS (MCQS)

25) Central nervous system is made up of ....?
 a) The brain and spinal cord
 b) Brain and heart
 c) Lung and Pharynx
 d) None of these

26) Which part of the brain controls breathing?

 a) Medulla oblongata
 b) Pineal gland
 c) Cerebellum
 d) None of these

Multiple Choice Questions (MCQs)

27) Controlling the intellectual ability is done by this part of the brain?
 a) Medulla
 b) Frontal lobe
 c) Temporal lobe
 d) None of these

28) Color of urine is yellow because of _____?
a) Haemoglobin
b) Uric Acid
c) Urochrome
d) Salt

MULTIPLE CHOICE QUESTIONS (MCQS)

29) The study of how body parts function called.....?

a) Anatomy
b) Histology
c) Physiology
d) None of these

30) The Functional unit of a human kidney is called?

a) Neuron
b) Protein
c) Nephron.
d) Cell

MCQ- ANSWERS

1(b)   2(d)   3(c)   4(a)   5 (d)   6(a)   7(d)   8 (d) 9(d) 10(d)

11(c) 12(c) 13(d) 14(a) 15(b) 16(c) 17(d) 18(a) 19(b)

20(d) 21(b) 22(b) 23(d) 24(c) 25(a) 26(a) 27(b)28(c)

29(c) 30(c)

DR. FANATOMY

FUN ACTIVITIES FOR MEDICOS

Appendix 1-Description of areas and body parts(1)

Term	Definition
Antebrachial	forearm
Axillary	armpit
Cardiac	heart
Cervical	neck
Cranial	head
Cutaneous	skin
Frontal	forehead

Appendix 1-Description of areas and body parts(2)

Term	Definition
Gastric	stomach
Glutea	buttocks
Hepatic	liver
Iliac	hip
Lumbar	back
Pectoral	chest
Pulmonary	lungs

Appendix 1-Description of areas and body parts(3)

Term	Definition
Renal	kidney
Sacral	base of spine
Orbital	eye
Umbilical	navel
Volar	palm
Buccal (oral)	mouth
Gluteal	buttocks

Appendix 2-Terminologies

Term	Definition
Anterior or ventral	Toward the front of the body
Posterior or dorsal	Toward the back of the body
Superior	A part above another part
Inferior	A part below another part
Medial	Toward the median plane of the body
Lateral	Away from the midline of the body
Proximal	Toward the point of attachment to the body

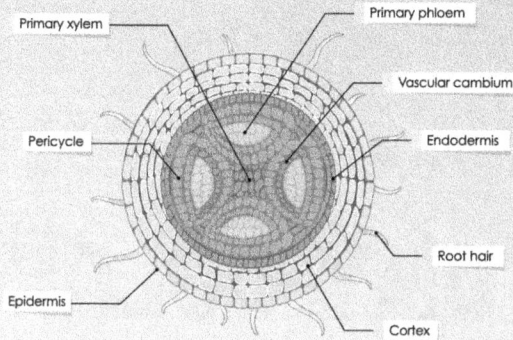

Primary xylem
Primary phloem
Vascular cambium
Pericycle
Endodermis
Epidermis
Root hair
Cortex

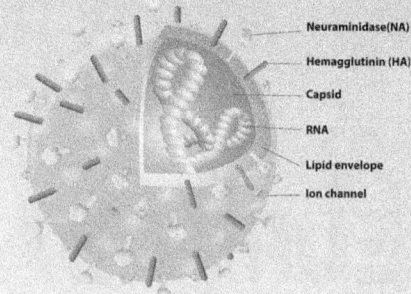

Neuraminidase(NA)
Hemagglutinin (HA)
Capsid
RNA
Lipid envelope
Ion channel

MICROBIOLOGY
COLORING BOOK WITH
FACTS & MCQS
(MULTIPLE CHOICE QUESTIONS)

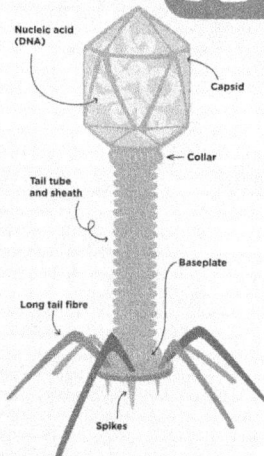

Nucleic acid (DNA)
Capsid
Collar
Tail tube and sheath
Baseplate
Long tail fibre
Spikes

DR. FANATOMY
FUN ACTIVITIES FOR MEDICOS

DR. FANATOMY

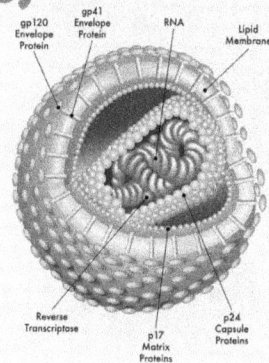

gp120 Envelope Protein
gp41 Envelope Protein
RNA
Lipid Membrane
Reverse Transcriptase
p17 Matrix Proteins
p24 Capsule Proteins

BOOK -2

NEUROANATOMY COLORING

BOOK

WITH MCQS

(MULTIPLE CHOICE QUESTIONS)

DR. FANATOMY

— FUN ACTIVITIES FOR MEDICOS —

CONTENTS

Introduction to the brain

Select different colors for each of the areas of the brain provided with a color-coding circle and use them to color in the coding circles and corresponding structures in the diagram.

①	Frontal lobe	⑪	Midbrain
②	Central sulcus	⑫	Pons
③	Parietal lobe	⑬	Medulla oblongata
④	Occipital lobe	⑭	Cerebellum
⑤	Lateral cerebral sulcus	⑮	Lateral ventricles
⑥	Cerebellum	⑯	Third ventricle
⑦	Temporal lobe	⑰	Fourth ventricle
⑧	Corpus callosum		
⑨	Thalamus		
⑩	Pituitary gland		

Notes:-

INTRODUCTION TO THE BRAIN

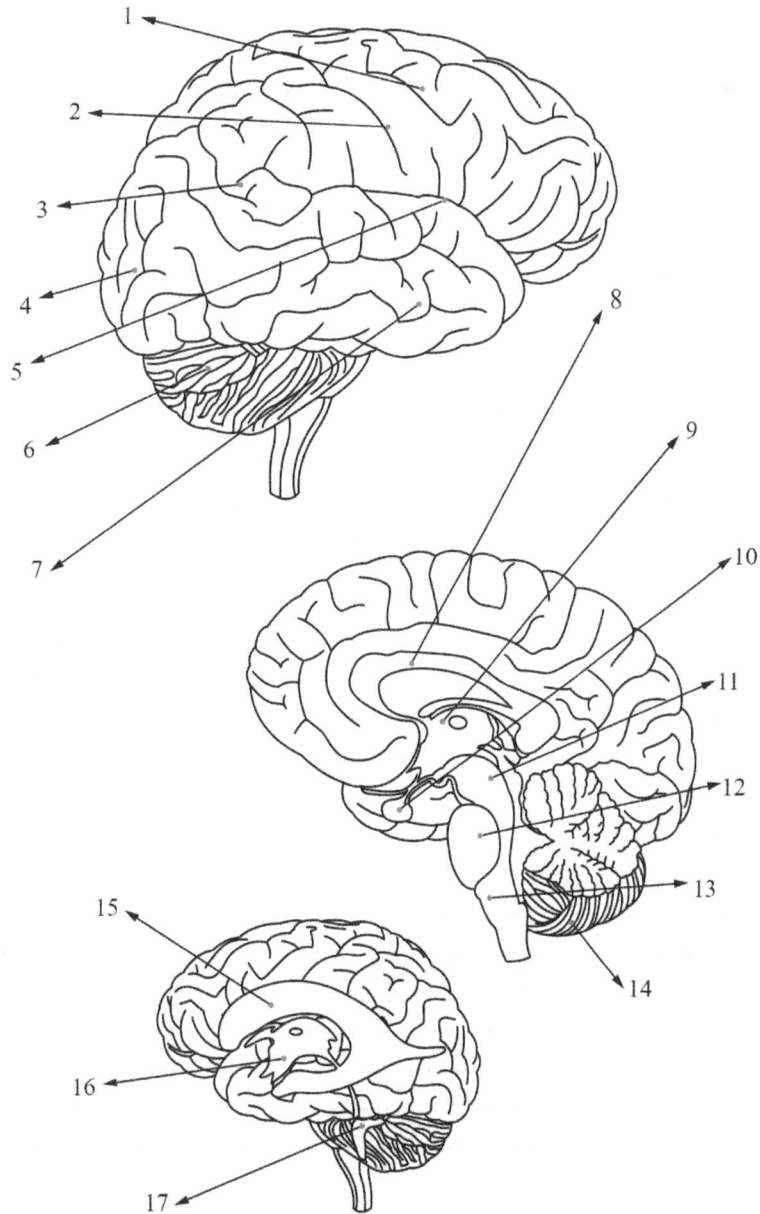

Lateral Views of the Brain

Select different colors for each of the areas of the brain provided with a color-coding circle and use them to color in the coding circles and corresponding structures in the diagram.

① Precentral sulcus	⑪ Superior temporal gyrus	㉑ Parietal lobe
② Superior frontal sulcus	⑫ Temporal lobe	㉒ Angular gyrus
③ Middle frontal gyrus	⑬ Middle temporal gyrus	㉓ Intraparietal sulcus
④ Superior frontal gyrus	⑭ Inferior temporal gyrus	㉔ Superior parietal lobule
⑤ Precentral gyrus	⑮ Inferior temporal sulcus	㉕ Postcentral sulcus
⑥ Inferior frontal sulcus	⑯ Superior temporal sulcus	㉖ Postcentral gyrus
⑦ Frontal pole	⑰ Supramarginal gyrus	㉗ Central sulcus
⑧ Frontal lobe	⑱ Occipital pole	
⑨ Inferior frontal gyrus	⑲ Occipital lobe	
⑩ Temporal pole	⑳ Inferior parietal lobule	

Notes:-

LATERAL VIEWS OF THE BRAIN

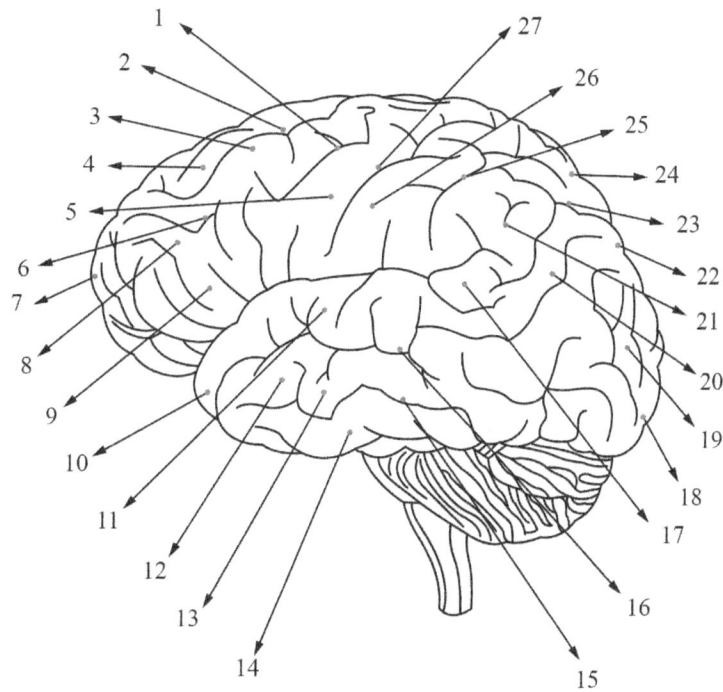

Superior View of the Brain

Select different colors for each of the areas of the brain provided with a color-coding circle and use them to color in the coding circles and corresponding structures in the diagram.

①	Left cerebral hemisphere	⑪	Right cerebral hemisphere	㉑	calcarine sulcus
②	Superior frontal gyrus	⑫	Superior frontal sulcus		
③	Middle frontal gyrus	⑬	Inferior frontal sulcus		
④	Inferior frontal gyrus	⑭	Longitudinal cerebral fissure		
⑤	Precentral gyrus	⑮	Precentral sulcus		
⑥	Postcentral gyrus	⑯	Lateral sulcus		
⑦	Supramarginal gyrus	⑰	Postcentral sulcus		
⑧	Angular gyrus	⑱	Intraparietal sulcus		
⑨	Superior parietal lobule	⑲	Preoccipital notch		
⑩	Occipital lobe	⑳	Parietooccipital sulcus		

Notes:-

SUPERIOR VIEW OF THE BRAIN

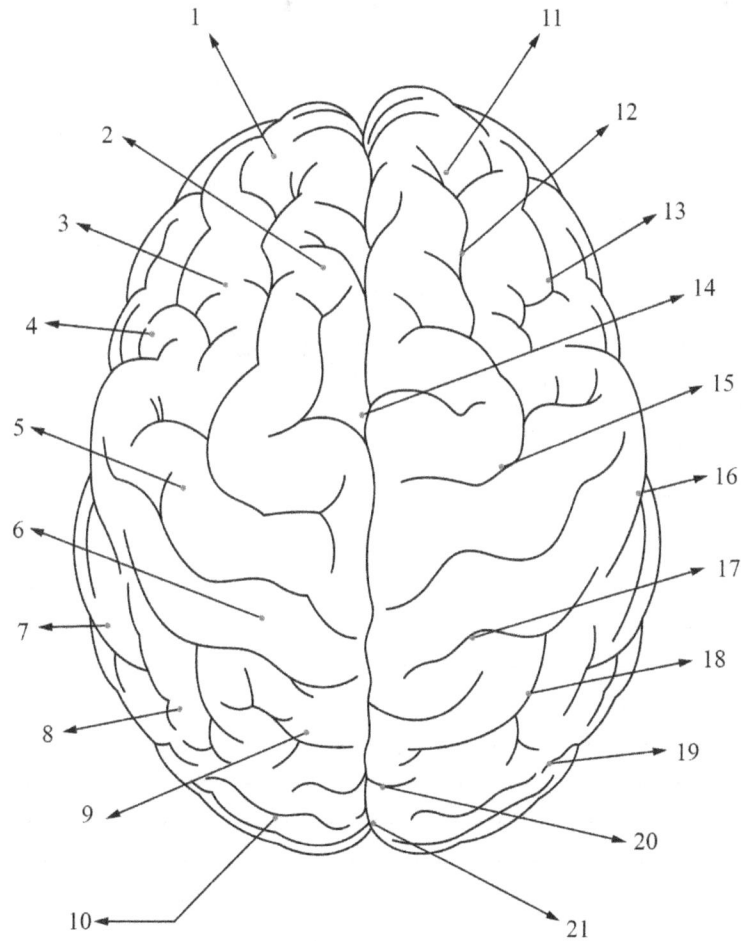

1
2
3
4
5
6
7
8
9
10
11
12
13
14
15
16
17
18
19
20
21

Medial View of the Brain

Select different colors for each of the areas of the brain provided with a color-coding circle and use them to color in the coding circles and corresponding structures in the diagram.

①	Paracentral lobule	⑪	Lamina terminalis	㉑	Superior medullary velum	㉛	Parietooccipital sulcus
②	Paracentral sulcus	⑫	Supraoptic recess	㉒	Cerebellum	㉜	Isthmus of cingulate gyrus
③	Cingulate gyrus	⑬	Optic chiasm	㉓	Cerebral aqueduct	㉝	Habenular commissure
④	Medial frontal gyrus	⑭	Infundibular recess	㉔	Inferior colliculus	㉞	Precuneus
⑤	Corpus callosum	⑮	Pituitary gland	㉕	Quadrigeminal plate	㉟	Choroid Plexus of third ventricle
⑥	Sulcus of corpus callosum	⑯	Mammillary bodies	㉖	Superior colliculus	㊱	Marginal sulcus
⑦	Septum pellucidum	⑰	Midbrain tegmentum	㉗	Calcarine sulcus	㊲	Central sulcus
⑧	Fornix	⑱	Pons	㉘	Posterior commissure		
⑨	Interthalamic adhesion	⑲	Medulla oblongata	㉙	Pineal gland		
⑩	Anterior commissure	⑳	Fourth ventricle	㉚	Cuneus		

Notes:-

MEDIAL VIEW OF THE BRAIN

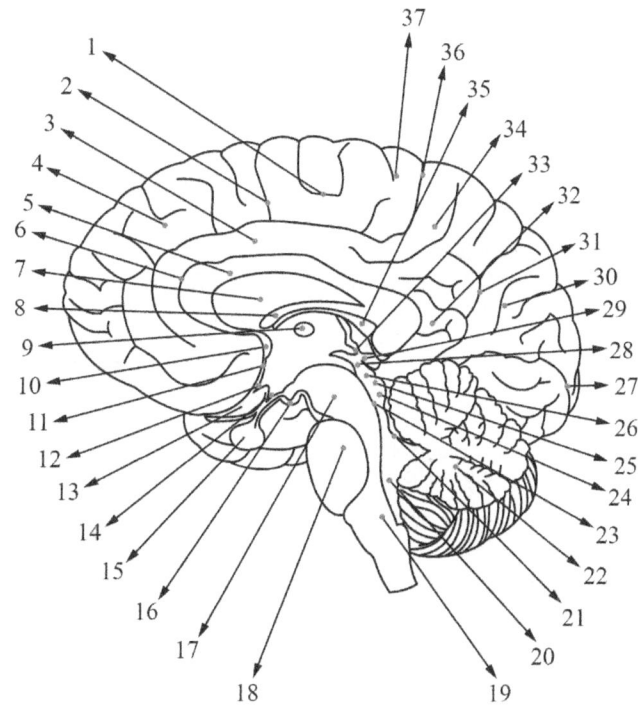

Basal View of the Brain

Select different colors for each of the areas of the brain provided with a color-coding circle and use them to color in the coding circles and corresponding structures in the diagram.

№		№		№		№	
①	Staright gyrus	⑪	Lateral geniculate nucleus	㉑	Occipital lobe	㉛	Inferior temporal gyrus
②	Olfactory sulcus	⑫	Red nucleus	㉒	Calcarine sulcus	㉜	Uncus
③	Orbital gyri	⑬	Occipitotemporal sulcus	㉓	Lingual gyrus	㉝	Optic tract
④	Orbital sulci	⑭	Medial geniculate nucleus	㉔	Isthmus of cingulate gyrus	㉞	Optic chiasm
⑤	Pituitary gland	⑮	Collateral sulcus	㉕	Splenium of corpus callosum	㉟	Optic nerve
⑥	Lateral sulcus	⑯	Pulvinar of thalamus	㉖	Lateral occipitotemporal gyrus	㊱	Temporal pole
⑦	Anterior perforated substance	⑰	Cerebral aqueduct	㉗	Substanstia nigra	㊲	Olfactory tract
⑧	Rhinal sulcus	⑱	Media occipitotemporal gyrus	㉘	Cerebral peduncle	㊳	Genu of corpus callosum
⑨	Mammillary bodies	⑲	Cuneus	㉙	Inferior temporal sulcus	㊴	Olfactory bulb
⑩	Posterior perforated substance	⑳	Longitudinal cerebral fissure	㉚	Parahippocampal gyrus	㊵	Frontal pole

Notes:-

BASAL VIEW OF THE BRAIN

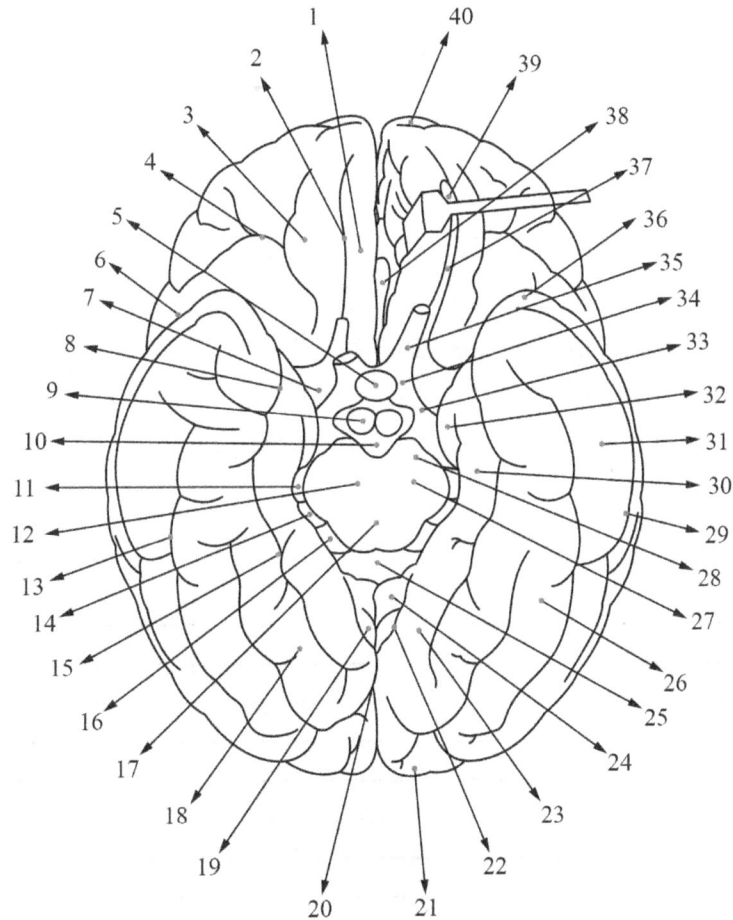

Brodmann areas

Select different colors for each of the areas of the brain provided with a color-coding circle and use them to color in the coding circles and corresponding structures in the diagram.

① Brodmann area 6	⑪ Brodmann area 43	㉑ Brodmann area 35	㉛ Brodmann area 41
② Brodmann area 8	⑫ Brodmann area 38	㉒ Brodmann area 28	㉜ Brodmann area 18
③ Brodmann area 4	⑬ Brodmann area 52	㉓ Brodmann area 34	㉝ Brodmann area 17
④ Brodmann area 9	⑭ Brodmann area 21	㉔ Brodmann area 25	㉞ Brodmann area 39
⑤ Brodmann area 46	⑮ Brodmann area 20	㉕ Brodmann area 32	㉟ Brodmann area 40
⑥ Brodmann area 10	⑯ Brodmann area 31	㉖ Brodmann area 33	㊱ Brodmann area 19
⑦ Brodmann area 44	⑰ Brodmann area 23	㉗ Brodmann area 24	㊲ Brodmann area 2
⑧ Brodmann area 45	⑱ Brodmann area 30	㉘ Brodmann area 22	㊳ Brodmann area 7
⑨ Brodmann area 11	⑲ Brodmann area 36	㉙ Brodmann area 42	㊴ Brodmann area 1
⑩ Brodmann area 47	⑳ Brodmann area 27	㉚ Brodmann area 37	㊵ Brodmann area 5
㊶ Brodmann area3			

Notes:-

BRODMANN AREAS

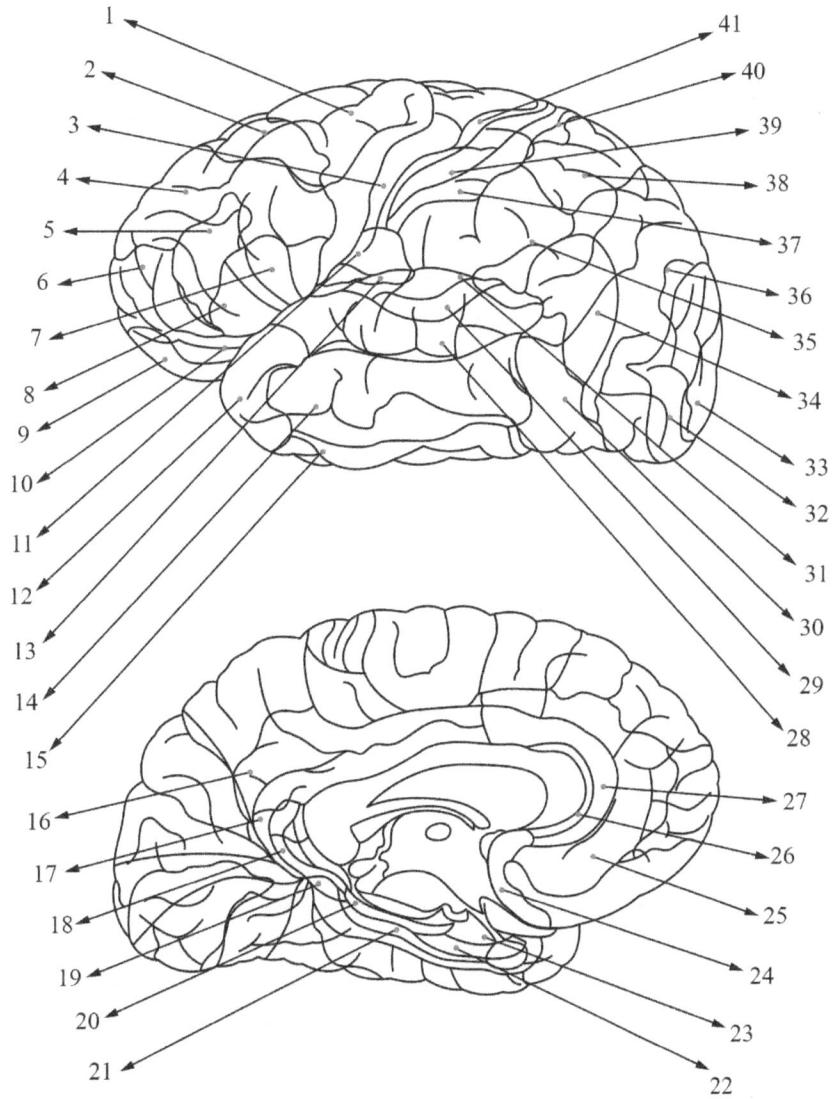

Thalamus

Select different colors for each of the areas of the brain provided with a color-coding circle and use them to color in the coding circles and corresponding structures in the diagram.

①	Corpus callosum	⑪	Longitudinal cerebral fissure	㉑	Fornix
②	Head of the caudate nucleus	⑫	Calcarine sulcus	㉒	Septum pellucidum
③	Internal capsule	⑬	Calcar avis		
④	Hippocampus	⑭	Inferior colliculus		
⑤	Third ventricle	⑮	Medial geniculate nucleus		
⑥	Fimbria of the hippocampus	⑯	Lateral geniculate nucleus		
⑦	Superior colliculus	⑰	Pulvinar of thalamus		
⑧	Lateral ventricle	⑱	Stria medullaris of thalamus		
⑨	Pineal gland	⑲	Thalamus		
⑩	Cerebellum	⑳	choroid plexus		

Notes:-

THALAMUS

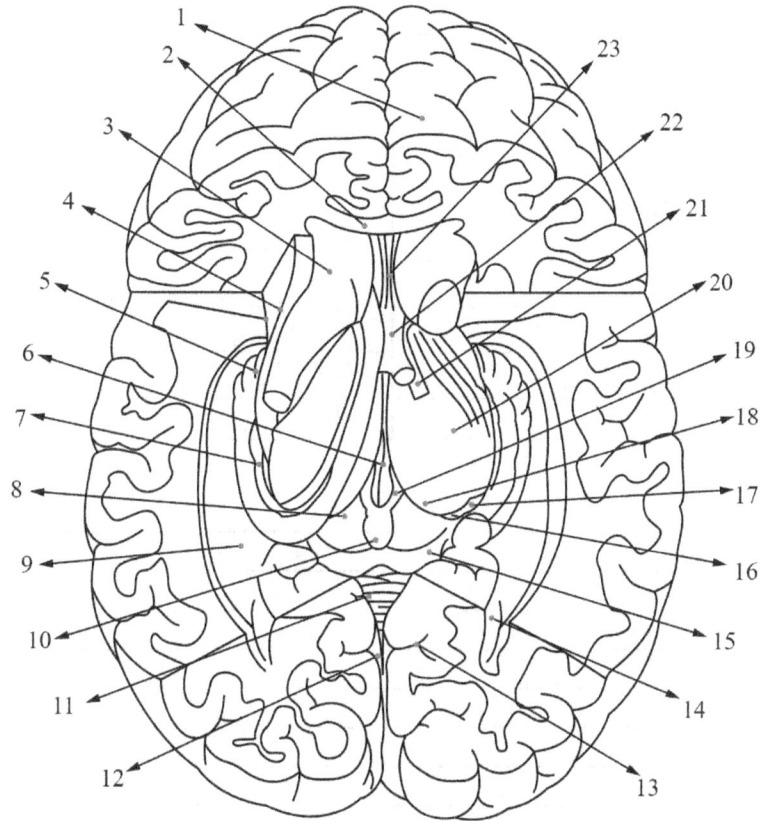

Hippocampus and Fornix

Select different colors for each of the areas of the brain provided with a color-coding circle and use them to color in the coding circles and corresponding structures in the diagram.

①	Thalamus	⑪	Dentate gyrus
②	Septum pellucidum	⑫	Fimbria of the hippocampus
③	Corpus callosum	⑬	Temporal horn of the lateral ventricle
④	Interthalamic adhesion	⑭	Lateral ventricle
⑤	Hypothalamus sulcus	⑮	Posterior horn of the lateral ventricle
⑥	Anterior commissure	⑯	Red nucleus
⑦	Tegmentum	⑰	Quadrigeminal plate
⑧	Mammillary bodies	⑱	Pineal gland
⑨	Substantia nigra	⑲	Stria medullaris of thalamus
⑩	Hippocampus	⑳	Fornix

Notes:-

HIPPOCAMPUS AND FORNIX

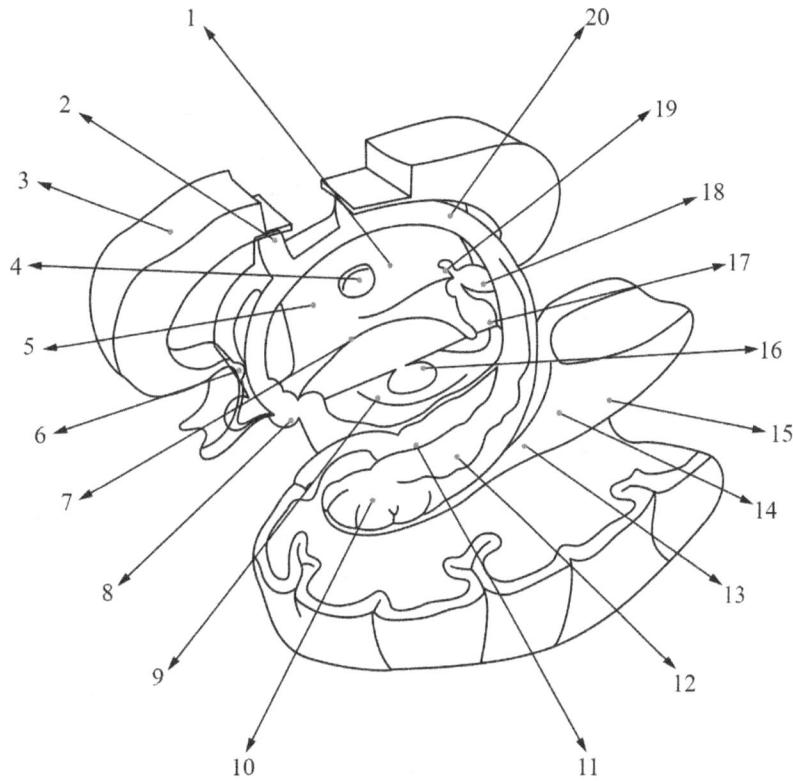

Thalamic nuclei

Select different colors for each of the areas of the brain provided with a color-coding circle and use them to color in the coding circles and corresponding structures in the diagram.

①	Anterior nuclei of thalamus	⑪	Pulvinar of thalamus
②	Lateral dorsal nucleus	⑫	Intralaminar nuclei of thalamus
③	Thalamic reticular nucleus	⑬	Medial medullary lamina
④	Ventral anterior nucleus	⑭	Midline nuclear group
⑤	Ventral lateral nucleus	⑮	Lateral thalamic nuclei
⑥	Lateral posterior nucleus	⑯	Medial dorsal nucleus
⑦	Ventral posteromedial nucleus		
⑧	Ventral posterolateral nucleus of thalamus		
⑨	Lateral geniculate nucleus		
⑩	Medial geniculate nucleus		

Notes:-

THALAMIC NUCLEI

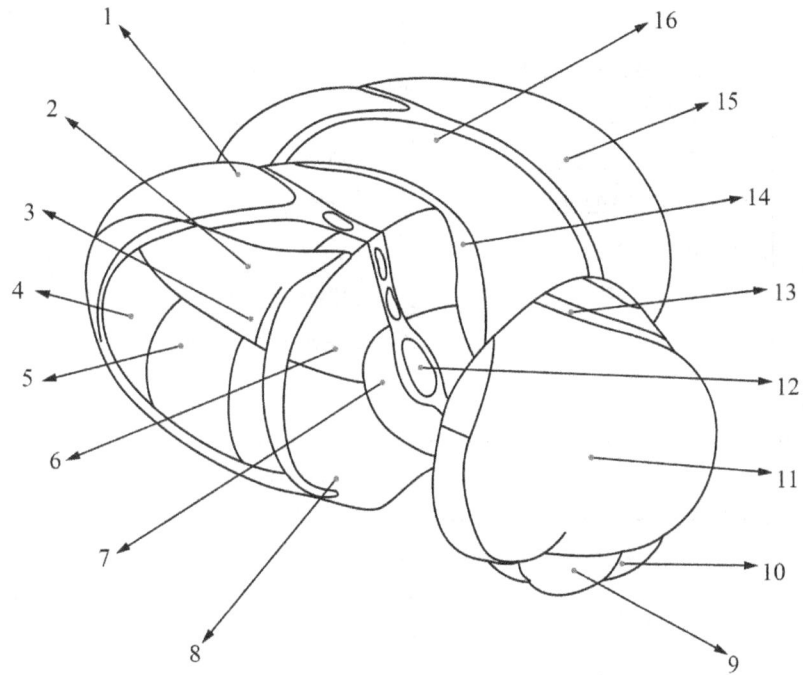

Hypothalamus

Select different colors for each of the areas of the brain provided with a color-coding circle and use them to color in the coding circles and corresponding structures in the diagram.

①	Fornix	⑪	Supraopticohypophyseal tract
②	Paraventricular nucleus	⑫	Mammillary complex
③	Dorsomedial hypothalamic nucleus	⑬	Red nucleus
④	Anterior hypothalamic nucleus	⑭	Intercalated nucleus
⑤	Lateral preoptic nucleus	⑮	Descending hypothalamic connection
⑥	Medial preoptic nucleus	⑯	Dorsal longitudinal fasciculus
⑦	Suprachiasmatic nucleus	⑰	Lateral hypothalamic area
⑧	Supraoptic nucleus	⑱	Mammillothalamic tract
⑨	Ventromedial nucleus	⑲	Preventricular nucleus
⑩	Arcuate nucleus	⑳	Posterior hypothalamic area

Notes:-

HYPOTHALAMUS

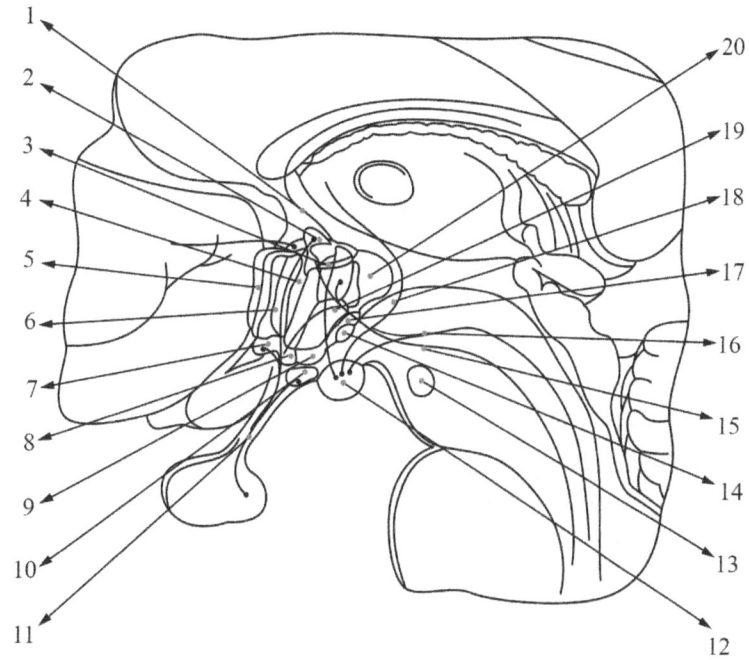

Pituitary gland

Select different colors for each of the areas of the brain provided with a color-coding circle and use them to color in the coding circles and corresponding structures in the diagram.

①	Lamina terminalis	⑪	Fibrous trabeculae of adenohypophysis
②	Optic chiasm	⑫	Neurohypophysis
③	Pars tuberalis of hypophysis	⑬	Mammillary body
④	Infundibular stalk	⑭	Median eminence of hypothalamus
⑤	Adenohypophysis		
⑥	Sella turcica		
⑦	Pars sistalis of hypophysis		
⑧	Hypophyseal cleft		
⑨	Pars intermedia of hypophysis		
⑩	Pars nervosa of hypophysis		

Notes:-

PITUITARY GLAND

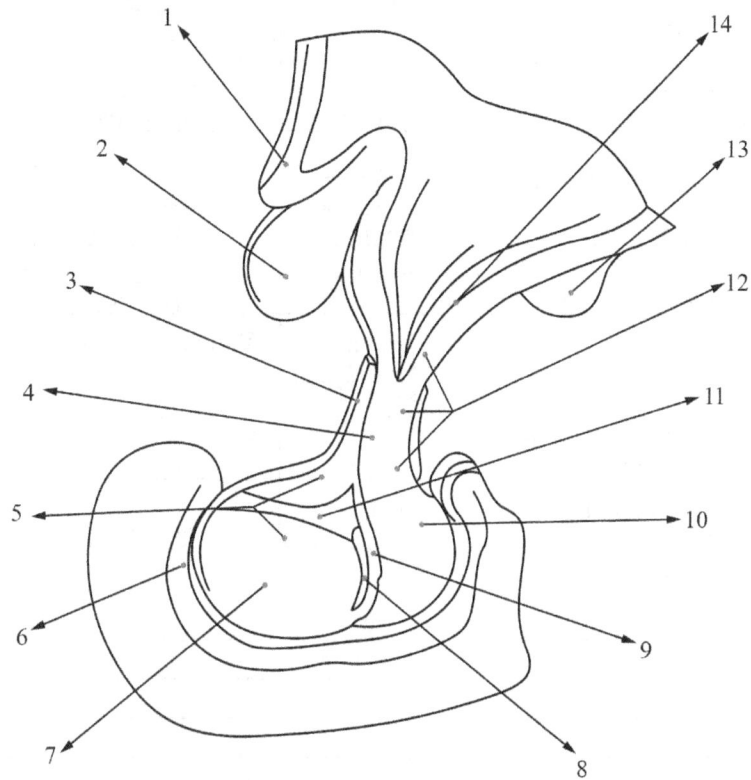

Hypophyseal portal system

Select different colors for each of the areas of the brain provided with a color-coding circle and use them to color in the coding circles and corresponding structures in the diagram.

① Lamina terminalis	⑪ Pars anterior of adenohypophysis	㉑ Neurohypophysis
② Optic chiasm	⑫ Fibrous trabeculae of adenohypophysis	㉒ Infundibular stem of neurohypophysis
③ Superior hypophyseal artery	⑬ Secondary capillary plexus of hypophyseal system	㉓ Primary capillary plexus of hypophyseal portal system
④ Optic nerve	⑭ Hypophyseal cleft	㉔ Mammillary bodies
⑤ Hypothalamic vessels	⑮ Pars Intermedia of adenohypophysis	㉕ Median eminence of hypothalamus
⑥ Long hypophyseal portal veins	⑯ Inferior hypophyseal artery	㉖ Arcuate nucleus of the hypothalamus
⑦ Infundibulum of pituitary gland	⑰ Capillary plexus of neurohypophysis	㉗ Supraopticohypophysial-Paraventriculohypophysial tract
⑧ Pars tuberalis of adenohypophysis	⑱ Pars posterior of neurohypophysis	㉘ Hypothalamus
⑨ Adenohypophysis	⑲ Hypophyseal vein	㉙ Supraoptic nucleus
⑩ Artery of trabecula of pituitary gland	⑳ Short hypophyseal portal vein	㉚ Paraventricular nucleus

Notes:-

HYPOPHYSEAL PORTAL SYSTEM

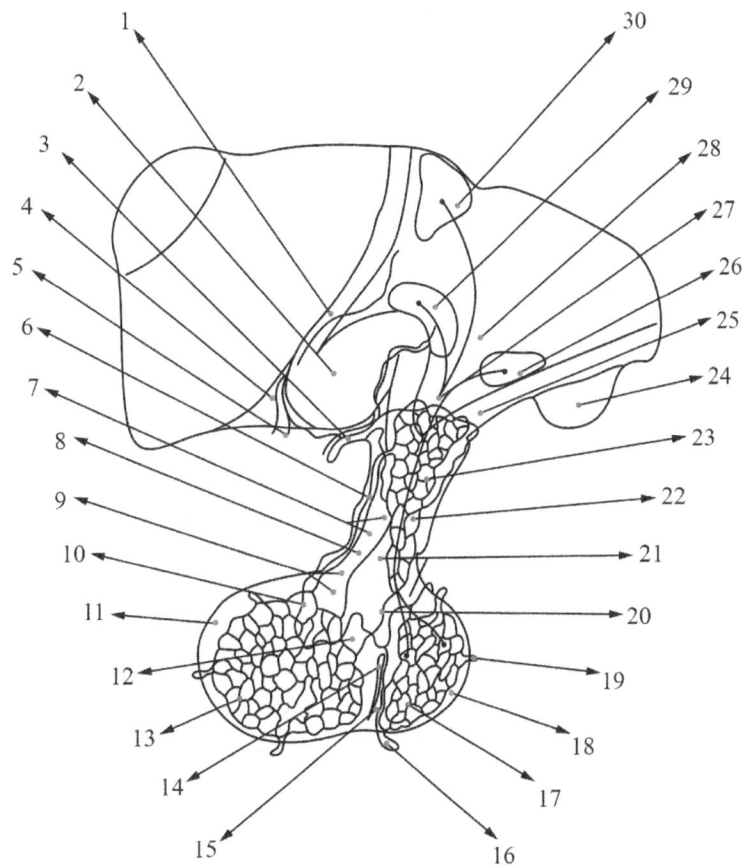

1

2

3

4

5

6

7

8

9

10

11

12

13

14

15

16

17

18

19

20

21

22

23

24

25

26

27

28

29

30

Basal Ganglia

Select different colors for each of the areas of the brain provided with a color-coding circle and use them to color in the coding circles and corresponding structures in the diagram.

①	Body	⑪	Thalamus
②	Tail	⑫	Terminal stria
③	Head		
④	Nucleus accumbens		
⑤	Putamen		
⑥	Lateral segment		
⑦	Medial segment		
⑧	Amygdaloid body		
⑨	Substantia nigra		
⑩	Subthalamic nucleus		

Notes:-

BASAL GANGLIA

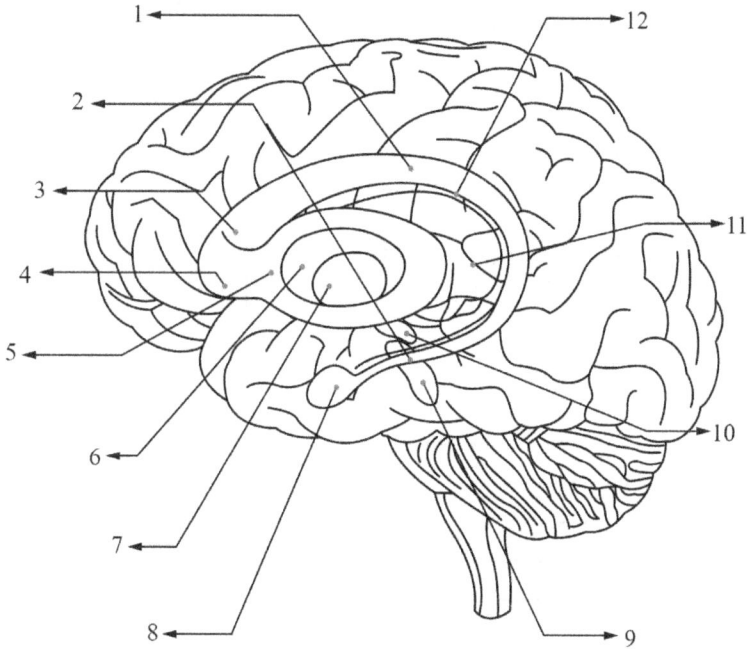

Coronal Section of the Brain at the thalamus Level

Select different colors for each of the areas of the brain provided with a color-coding circle and use them to color in the coding circles and corresponding structures in the diagram.

①	Body of corpus callosum	⑪	Substantia nigra	㉑	Central part of the lateral ventricle
②	Choroid plexus	⑫	Hippocampus	㉒	Body of fornix
③	Thalamus	⑬	Optic tract	㉓	Cingulate gyrus
④	Internal capsule	⑭	Tail of the caudate nucleus		
⑤	Extreme Capsule	⑮	Globus pallidus internal segment		
⑥	Putramen	⑯	External capsule		
⑦	Globus pallidus internal segment	⑰	Insular lobe		
⑧	Temporal horn of the lateral ventricle	⑱	Lenticular nucleus		
⑨	Subthalamic nucleus	⑲	Third ventricle		
⑩	Mammillary bodies	⑳	Body of caudate nucleus		

Notes:-

CORONAL SECTION OF THE BRAIN
AT THE THALAMUS LEVEL

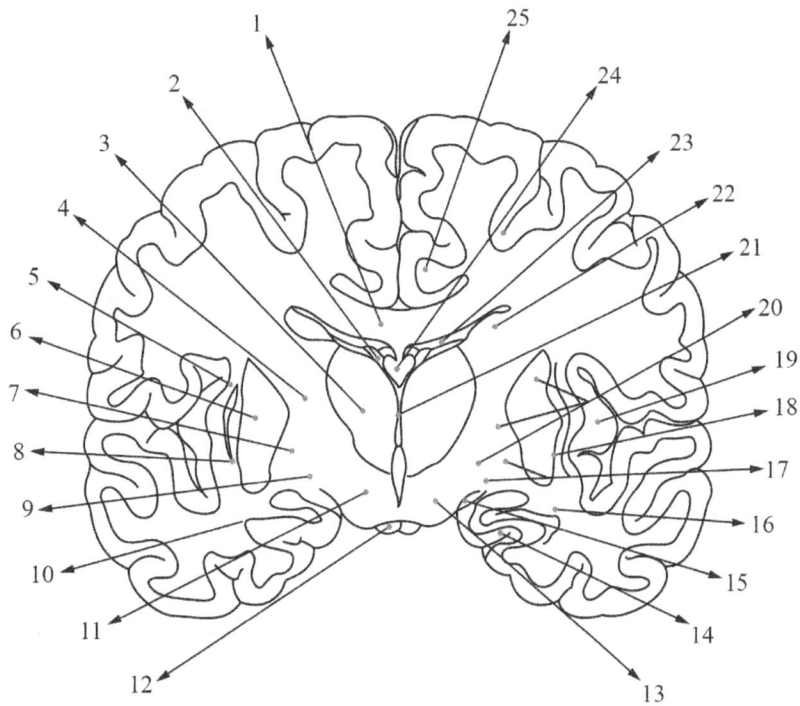

Horizontal Section of the Brain

Select different colors for each of the areas of the brain provided with a color-coding circle and use them to color in the coding circles and corresponding structures in the diagram.

№		№		№	
①	Anterior horn of the lateral ventricle	⑪	Splenium of corpus callosum	㉑	Third ventricle
②	Head of the caudate nucleus	⑫	Pineal gland	㉒	Claustrum
③	Column of fornix	⑬	Calcarine sulcus	㉓	Insular lobe
④	Extreme capsule	⑭	Cerebellum	㉔	Putamen
⑤	External capsule	⑮	Hippocampus	㉕	Globus pallidus
⑥	Interthalamic adhesion	⑯	Posterior horn of the lateral ventricle	㉖	Septum pellucidum
⑦	Internal capsule	⑰	Fimbria of the hippocampus	㉗	Genu of corpus callosum
⑧	Thalamus	⑱	Tail of the caudate nucleus	㉘	Cingulate gyrus
⑨	Crus of fronix	⑲	Habenula		
⑩	Choroid plexus	⑳	Retrolenticular part of the internal capsule		

Notes:-

HORIZONTAL SECTION OF THE BRAIN

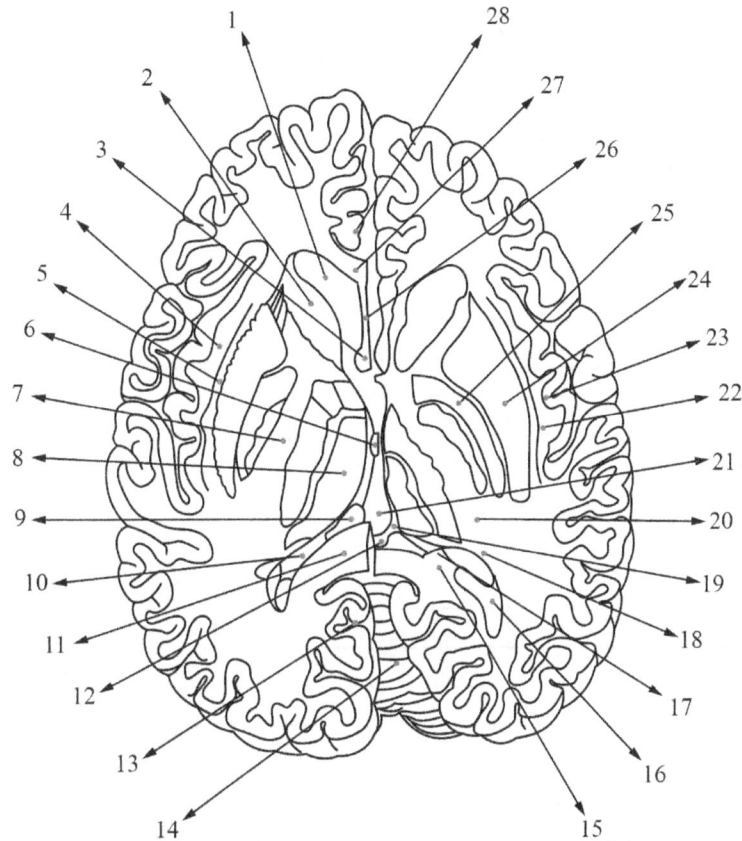

Cerebellum

Select different colors for each of the areas of the brain provided with a color-coding circle and use them to color in the coding circles and corresponding structures in the diagram.

① Culmen	⑪ Anterior lobe of cerebellum	㉑ Cerebellar tonsil
② Primary fissure of cerebellum	⑫ Quadrangular lobule	㉒ Pyramid of cerebellum
③ Declive	⑬ Central lobule	㉓ Uvula of vermis
④ Posterior lobe of cerebellum	⑭ Superior medullary velum	㉔ Inferior cerebellar peduncle
⑤ Superior of semilunar lobule	⑮ Inferior medullary velum	㉕ Nodule of vermis
⑥ Horizontal fissure of cerebellum	⑯ Middle cerebellar peduncle	㉖ Superior of cerebellar peduncle
⑦ Inferior semilunar lobule	⑰ Flocculonodular lobe	㉗ Fourth ventricle
⑧ Folium	⑱ Flocculus	㉘ Lingula of cerebellum
⑨ Simple lobule	⑲ Biventer lobule	
⑩ Vermis	⑳ Retrotonsillar fissure of cerebellum	

Notes:-

CEREBELLUM

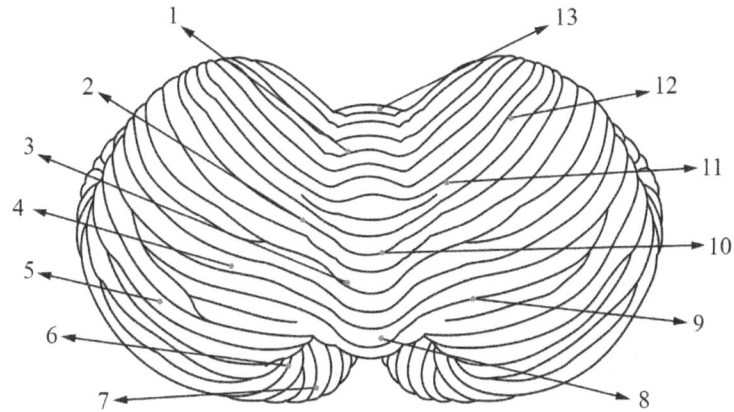

1
2
3
4
5
6
7
13
12
11
10
9
8

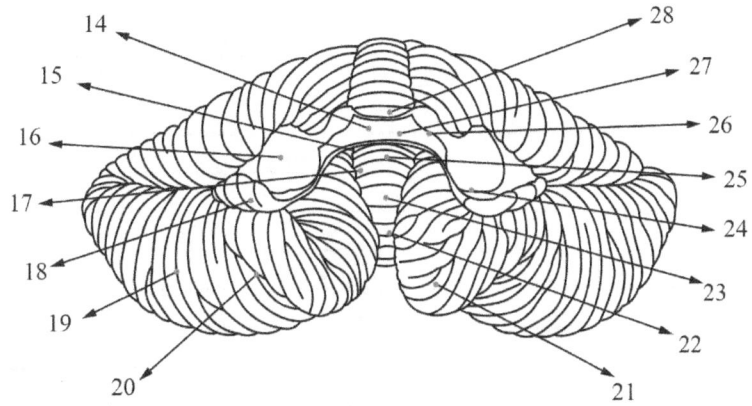

14
15
16
17
18
19
20
28
27
26
25
24
23
22
21

Cerebellar Nuclei

Select different colors for each of the areas of the brain provided with a color-coding circle and use them to color in the coding circles and corresponding structures in the diagram.

		⑪	
①			Decussation of superior cerebellar peduncle
	Superior medullary velum		
②	Superior cerebellar peduncle		
③	Lingula of cerebellum		
④	Emboliform nuclei		
⑤	Fastigial nuclei		
⑥	Vermis		
⑦	Dentate nucleus		
⑧	Globose nuclei		
⑨	Fourth ventricle		
⑩	Medial longitudinal fasciculus		

Notes:-

CEREBELLAR NUCLEI

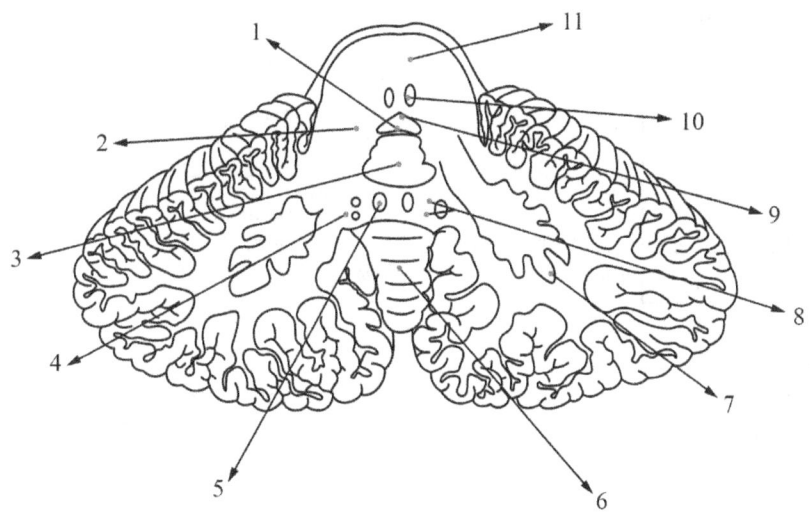

Brainstem

Select different colors for each of the areas of the brain provided with a color-coding circle and use them to color in the coding circles and corresponding structures in the diagram.

① Third Ventricle	⑪ Locus coeruleus	㉑ Lateral funiculus	㉛ Trochlear nerve
② Taenia thalami	⑫ Inferior cerebellar peduncle	㉒ Cuneate tubercle	㉜ Lateral geniculate nucleus
③ Choroid line	⑬ Medial eminence	㉓ Gracile tubercle	㉝ Medial geniculate nucleus
④ Caudate nucleus	⑭ Vestibular area	㉔ Obex	㉞ Habenular trigone
⑤ Pineal gland	⑮ Hypoglossal trigone	㉕ Cerebellum	㉟ Thalamus
⑥ Superior colliculus	⑯ Trigeminal tubercle	㉖ Medullary striae of fourth ventricle	㊱ Terminal stria
⑦ Cerebral peduncle	⑰ Vagal trigone	㉗ Dentate nucleus	㊲ Anterior thalamic tubercle
⑧ Inferior colliculus	⑱ Cuneate fasciculus	㉘ Cerebellar cortex	
⑨ Frenulum of siperior medullary velum	⑲ Gracile fasciculus	㉙ Superior Cerebellar peduncle	
⑩ Superior medullary velum	⑳ Posterior median sulcus	㉚ Middle cerebellar peduncle	

Notes:-

BRAINSTEM

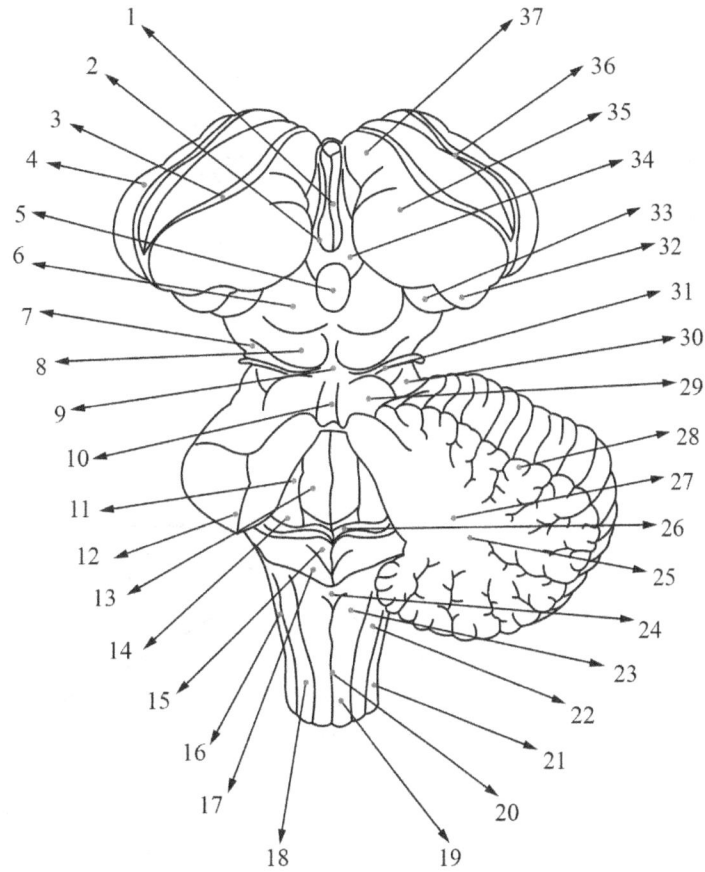

Anterior View of the Brainstem

Select different colors for each of the areas of the brain provided with a color-coding circle and use them to color in the coding circles and corresponding structures in the diagram.

① Olfactory tract	⑪ Vagus nerve	㉑ Facial nerve
② Intermediate hypothalamic area	⑫ Choroid plexus of fourth ventricle	㉒ Posterior perforated substance
③ Oculomotor nerve	⑬ Medullary pyramids	㉓ Trochlear nerve
④ Cerebral peduncle	⑭ Pyramidal decussation	㉔ Temporal lobe
⑤ Optic tract	⑮ Rootlets of the ventral root	㉕ Mammillary bodies
⑥ Pons	⑯ Cervical part of spinal cord	㉖ Infundibulum of pituitary gland
⑦ Trigeminal nerve	⑰ Accessory nerve	㉗ Anterior perforated substance
⑧ Abducens nerve	⑱ Hypoglossal nerve	㉘ Optic chiasm
⑨ Vestibulocochlear nerve	⑲ Glossophyarngeal nerve	㉙ Optic nerve
⑩ Olivary nuclei	⑳ Flocculus	

Notes:-

ANTERIOR VIEW OF THE BRAINSTEM

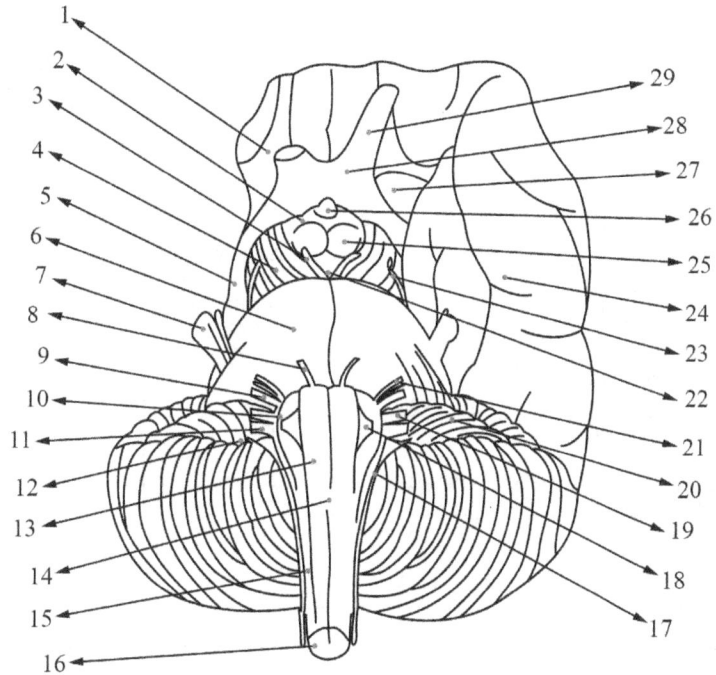

Cranial Nerve Nuclei

Select different colors for each of the areas of the brain provided with a color-coding circle and use them to color in the coding circles and corresponding structures in the diagram.

① Red Nucleus	⑪ Abducens nucleus	㉑ Mesencephalon	㉛ Solitary nucleus and tract
② Oculomotor nerve	⑫ Nucleus of facial nerve	㉒ Cerebellum	㉜ Vestibular nuclei
③ Pons	⑬ Superior salivatory nucleus	㉓ Medulla oblongata	㉝ Anterior cochlear nucleus
④ Abducens nerve	⑭ Inferior salivatory nucleus	㉔ Olivary nuclei	㉞ Posterior cochlear nucleus
⑤ Hypoglossal nerve	⑮ Facial nerve	㉕ Superior colliculus	㉟ Vestibulocochlear nerve
⑥ Accessory oculomotor nucleus	⑯ Nucleus ambiguus	㉖ Trochlear nucleus	㊱ Glossopharyngeal nerve
⑦ Oculomotor nucleus	⑰ Vagus nerve	㉗ Lateral geniculate nucleus	㊲ Spinal nucleus and tract of trigeminal nerve
⑧ Trochlear nerve	⑱ Hypoglossal nucleus	㉘ Mesencephalic nucleus of trigeminal nerve	㊳ Dorsal nucleus of vagus nerve
⑨ Trigeminal motor nucleus	⑲ Accessory nerve	㉙ Trigeminal ganglion	
⑩ Trigeminal nerve	⑳ Spinal accessory nucleus	㉚ Principal (sensory) nucleus of trigeminal nerve	

Notes:-

CRANIAL NERVE NUCLEI

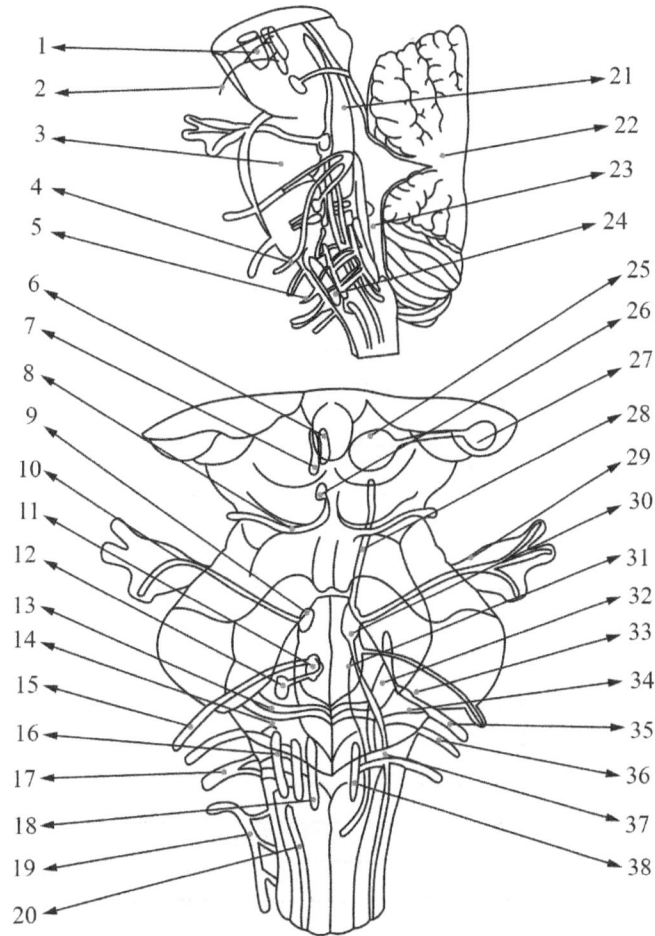

1
2
3
4
5
6
7
8
9
10
11
12
13
14
15
16
17
18
19
20

21
22
23
24
25
26
27
28
29
30
31
32
33
34
35
36
37
38

Medulla Oblongata:Hypoglossal Nerve Level

Select different colors for each of the areas of the brain provided with a color-coding circle and use them to color in the coding circles and corresponding structures in the diagram.

①	Nucleus solitary tract	⑪	Arcuate nucleus	㉑	Gracile nucleus
②	Solitary tract	⑫	Olivocerebellar tract		
③	Cuneate nucleus	⑬	Olivary nuclei		
④	Medial longitudinal fasciculus	⑭	Posterior accessory olivary nucleus		
⑤	Nucleus ambiguus	⑮	Lateral reticular nucleus		
⑥	Internal arcuate fibers	⑯	Spinal tract of trigeminal nerve		
⑦	Central tegmental tract	⑰	Spinal trigeminal nucleus		
⑧	Medial lemniscus	⑱	Hypoglossal nerve		
⑨	Superficial arcuate fibers	⑲	Hypoglossal nucleus		
⑩	Pyramidal tract	⑳	Dorsal nucleus of vagus nerve		

Notes:-

MEDULLA OBLONGATA: HYPOGLOSSAL NERVE LEVEL

Medulla Oblongata Vagus Nerve level

Select different colors for each of the areas of the brain provided with a color-coding circle and use them to color in the coding circles and corresponding structures in the diagram.

①	Medial vestibular nucleus	⑪	Inferior olivary nucleus
②	Posterior nucleus of vagus nerve	⑫	Raphe nuclei
③	Inferior cerebellar peduncle	⑬	Dorsal accessory olivary nucleus
④	Spinal trigeminal nucleus	⑭	Spinocerebellar tract
⑤	Medial longitudinal fasciculus	⑮	Nucleus ambiguus
⑥	Lateral reticular nucleus	⑯	Reticular formation
⑦	Spinothalamic tract	⑰	Spinal tract of trigeminal nerve
⑧	Medial accessory olivary nucleus	⑱	Cuneate nucleus
⑨	Medial lemniscus	⑲	Roller's nucleus
⑩	Pyramidal tract	⑳	Hypoglossal nucleus

Notes:-

MEDULLA OBLONGATA:
VAGUS NERVE LEVEL

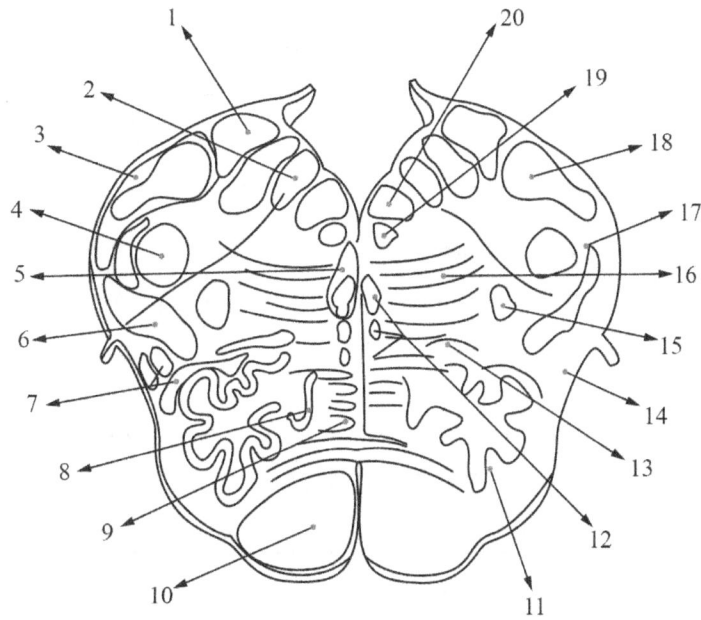

Ventricles of the brain

Select different colors for each of the areas of the brain provided with a color-coding circle and use them to color in the coding circles and corresponding structures in the diagram.

①	Left Lateral ventricle	⑪	Fourth ventricle
②	Intraventricular foramen	⑫	Cerebral aqueduct
③	Anterior horn of the lateral ventricle	⑬	Posterior horn of the lateral ventricle
④	Right lateral ventricle	⑭	Collateral trigone
⑤	Third ventricle	⑮	Recess of pineal gland
⑥	Supraoptic recess	⑯	Suprapineal recess
⑦	Infundibular recess	⑰	Central part of the lateral ventricle
⑧	Temporal horn of the lateral ventricle		
⑨	Central canal of spinal cord		
⑩	Lateral recess		

Notes:-

VENTRICLES OF THE BRAIN

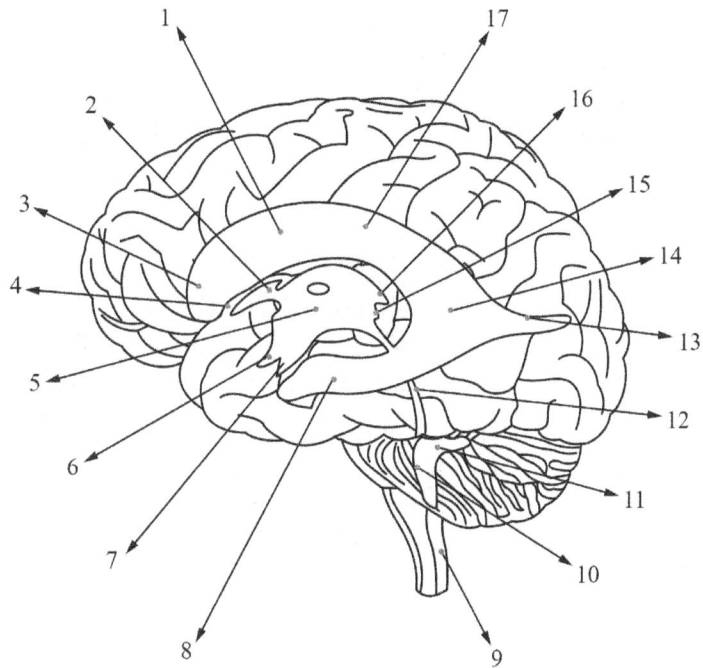

Arteries of the Brain

Select different colors for each of the areas of the brain provided with a color-coding circle and use them to color in the coding circles and corresponding structures in the diagram.

①	Artery of precentral sulcus	⑪	Cingular branches	㉑	Right posterior cerebral artery
②	Posterior parietal artery	⑫	Pericallosal artery	㉒	Anterior temporal artery
③	Anterior parietal artery	⑬	Callosomarginal artery	㉓	Posterior temporal artery
④	Branch of middle cerebral artery to angular gyrus	⑭	Frontal branches of middle cerebral artery	㉔	Medial occipital artery
⑤	Middle temporal branch of middle cerebral artery	⑮	Polar frontal artery	㉕	Calcarine branch of medial occipital artery
⑥	Superior and inferior terminal branches of middle cerebral artery	⑯	Right anterior cerebral artery	㉖	Parietooccipital branch of medial occipital artery
⑦	Middle cerebral artery	⑰	Medial frontobasal artery	㉗	Dorsal branch to corpus callosum
⑧	Lateral frontobasal artery	⑱	Anterior communicating artery	㉘	Precuneal branches of pericallosal artery
⑨	Artery of prefontal sulcus	⑲	Right internal carotid artery	㉙	Paracentral artery
⑩	Artery of central sulcus	⑳	Posterior communicating artery		

Notes:-

ARTERIES OF THE BRAIN

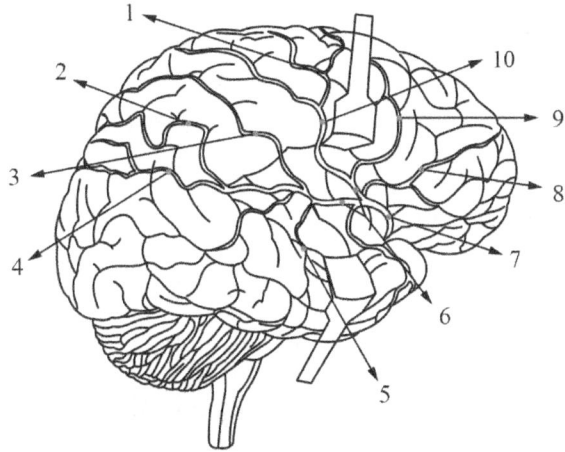

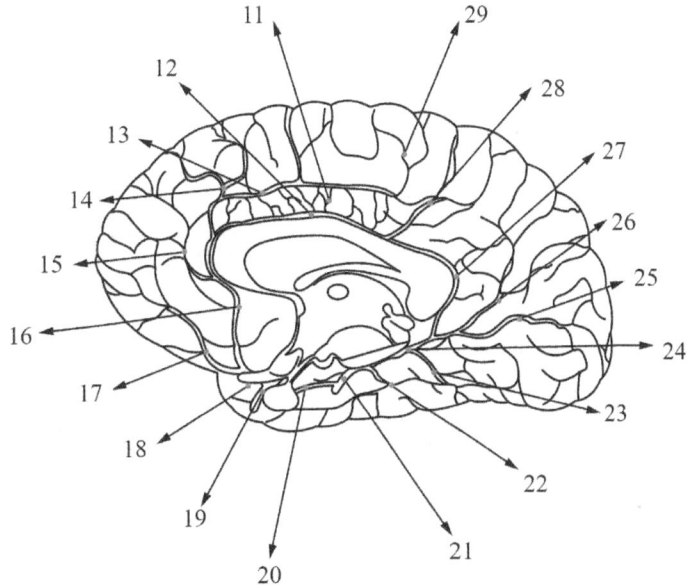

Arteries of the brain II

Select different colors for each of the areas of the brain provided with a color-coding circle and use them to color in the coding circles and corresponding structures in the diagram.

①	Anterior cerebral artery	⑪	Anterior spinal artery
②	Lateral frontobasal artery	⑫	Anterior inferior cerebellar artery
③	Middle cerebral artery	⑬	Basilar artery
④	Artery of prefontal sulcus	⑭	Pontine arteries
⑤	Posterior communicating artery	⑮	Anterior choroidal artery
⑥	Superior cerebellar artery	⑯	Internal carotid artery
⑦	Posterior cerebral artery	⑰	Anterior communicating artery
⑧	Labyrinthine artery	⑱	Medial frontobasal artery
⑨	Vertebral artery		
⑩	Posterior inferior cerebellar artery		

Notes:-

ARTERIES OF THE BRAIN II

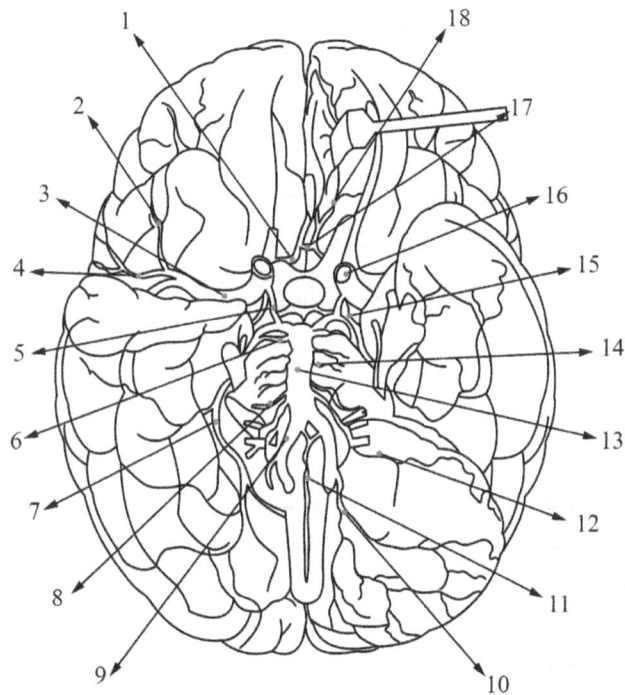

Superficial veins of the brain

Select different colors for each of the areas of the brain provided with a color-coding circle and use them to color in the coding circles and corresponding structures in the diagram.

①	Inferior sagittal sinus	⑪	Superficial middle cerebral vein	㉑	Superior anastomotic vein
②	Anterior cerebral vein	⑫	Superior petrosal sinus		
③	Basal vein	⑬	Inferior petrosal sinus		
④	Superior cerebellar vein	⑭	Petrosal vein		
⑤	Great cerebral vein	⑮	Internal jugular vein		
⑥	Internal occipital vein	⑯	Sigmoid sinus		
⑦	Straight sinus	⑰	Occipital sinus		
⑧	Posteromedian medullary vein	⑱	Confluence of sinuses		
⑨	Superior sagittal sinus	⑲	Transverse sinus		
⑩	Superior cerebral veins	⑳	Inferior anastomotic vein		

Notes:-

SUPERFICIAL VEINS OF THE BRAIN

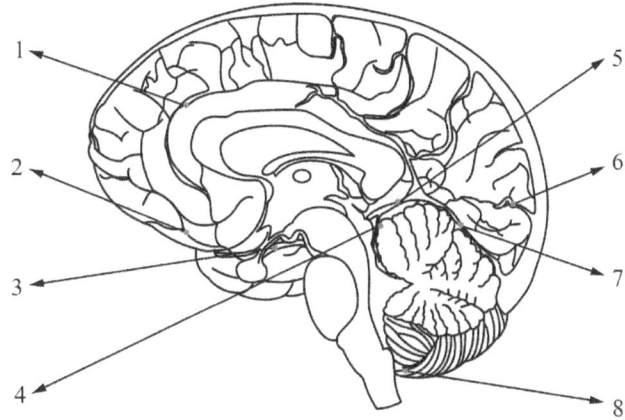

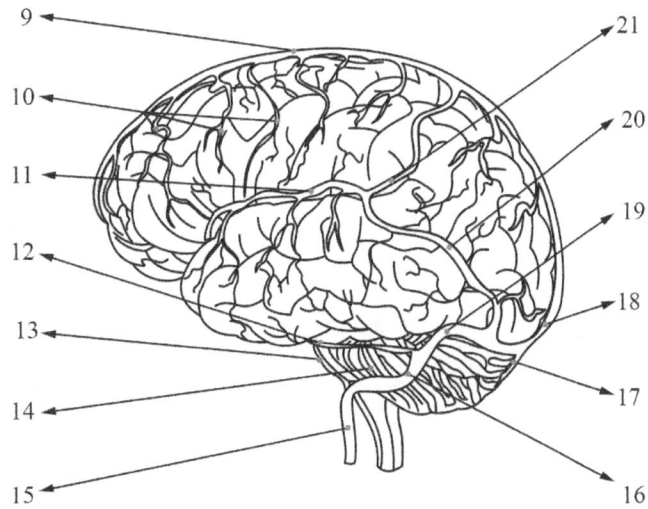

Superficial veins of the brain II

Select different colors for each of the areas of the brain provided with a color-coding circle and use them to color in the coding circles and corresponding structures in the diagram.

① Anterior cerebral vein	⑪ Straight sinus	㉑ Anterolateral medullary vein	㉛ Superficial middle cerebral vein
② Deep middle cerebral vein	⑫ Confluence of sinuses	㉒ Posteromedian medullary vein	㉜ Sphenoparietal sinus
③ Intrapeduncular vein	⑬ Anterior intercavernous sinus	㉓ Transverse sinus	㉝ Inferior opthalmic vein
④ Basal vein	⑭ Posterior intercavernous sinus	㉔ Cerebellar vein	㉞ Superior opthalmic vein
⑤ Posterior venous confluence	⑮ Pontomesencephalic vein	㉕ Sigmoid sinus	
⑥ Occipital sinus	⑯ Anteromedian pontine vein	㉖ Inferior petrosal sinus	
⑦ Anterior communicating vein	⑰ Superior petrosal veins	㉗ Superior cerebellar vein	
⑧ Inferior choroidal vein	⑱ Anterolateral pontine vein	㉘ Superior petrosal sinus	
⑨ Internal cerebral vein	⑲ Transverse medullary vein	㉙ Transverse pontine vein	
⑩ Great cerebral vein	⑳ Anteromedian medullary vein	㉚ Cavernous sinus	

Notes:-

SUPERFICIAL VEINS OF THE BRAIN II

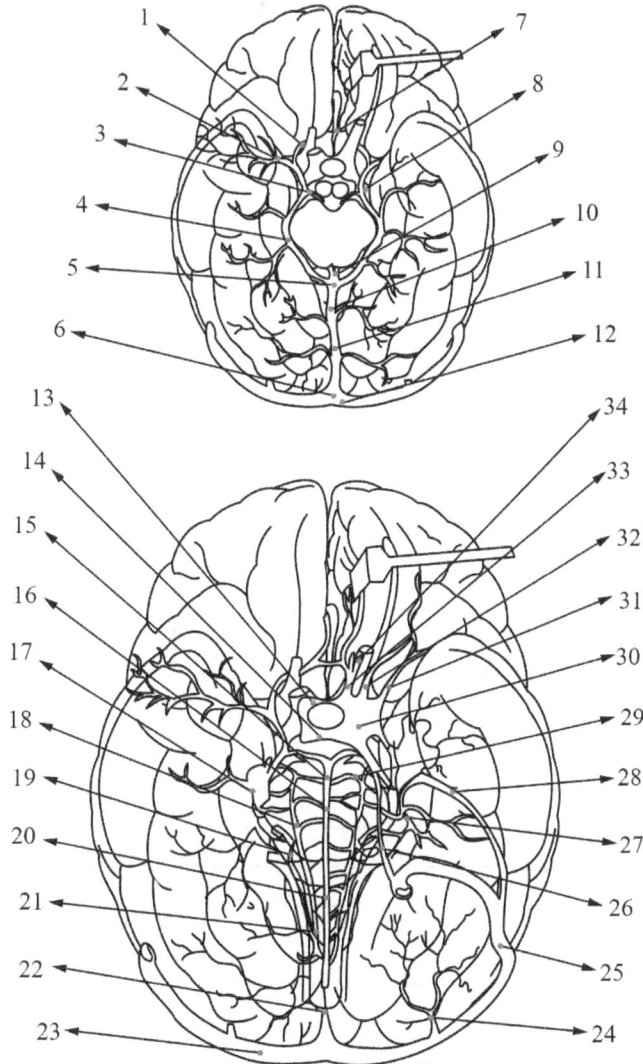

Meninges of the brain

Select different colors for each of the areas of the brain provided with a color-coding circle and use them to color in the coding circles and corresponding structures in the diagram.

①	Dura mater	
②	Middle meningeal artery	
③	Arachnoid	
④	Arachnoid granulation	
⑤	Confluence of sinuses	
⑥	Pia mater	
⑦	Superior sagittal sinus	
⑧	Branches of middle cerebral artery	
⑨	Superior cerebral veins	
⑩	Lateral lacunae of superior sagittal sinus	

Notes:-

MENINGES OF THE BRAIN

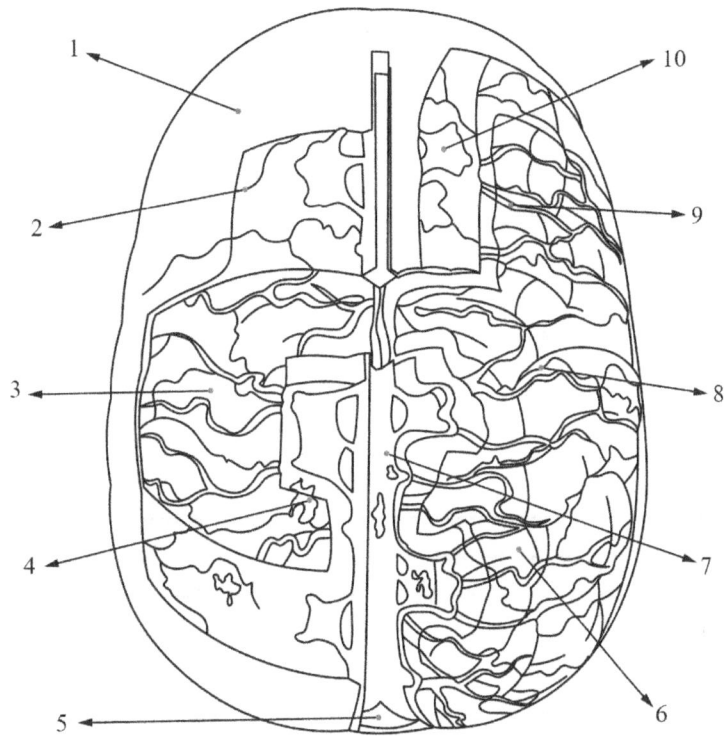

Arachnoid granulations

Select different colors for each of the areas of the brain provided with a color-coding circle and use them to color in the coding circles and corresponding structures in the diagram.

①	Periosteum	⑪	Arachnoid cap cells
②	Emissary veins	⑫	Arachnoid granulation
③	Skull	⑬	Pia mater
④	Subdural space	⑭	Meningeal layer of dura mater
⑤	Superior sagittal sinus	⑮	Periosteal layer of dura mater
⑥	Subarachnoid space	⑯	Diploic veins
⑦	Arachnoid mater		
⑧	Cerebral cortex		
⑨	Endothelium		
⑩	Arachnoid trabeculae		

Notes:-

ARACHNOID GRANULATIONS

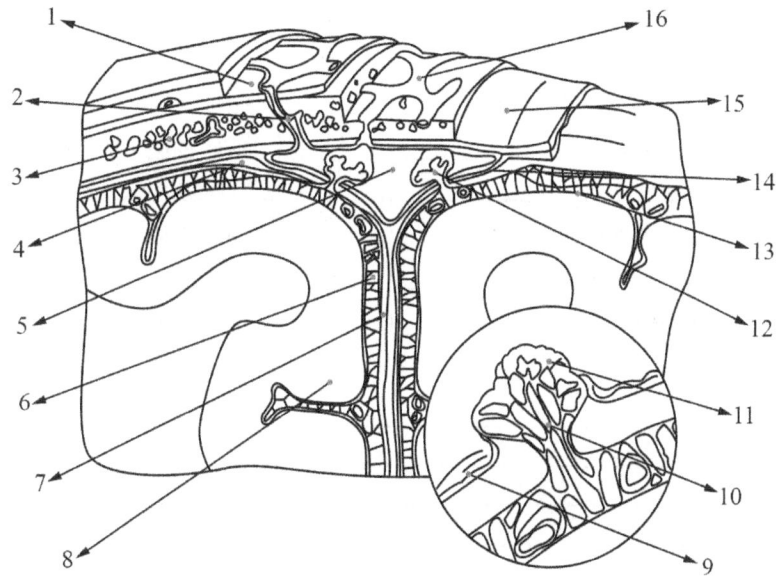

Dural venous sinuses

Select different colors for each of the areas of the brain provided with a color-coding circle and use them to color in the coding circles and corresponding structures in the diagram.

① Superficial middle cerebral vein	⑪ Superior sagittal sinus
② Sphenoparietal sinus	⑫ Confluence of sinuses
③ Internal carotid artery	⑬ Straight sinus
④ Cavernous sinus	⑭ Great cerebral vein
⑤ Posterior intercavernous sinus	⑮ Inferior petrosal sinus
⑥ Superior petrosal sinus	⑯ Basilar venous plexus
⑦ Sigmoid sinus	⑰ Trigeminal nerve
⑧ Inferior cerebral veins	⑱ Pituitary gland
⑨ Inferior sagittal sinus	⑲ Optic chiasm
⑩ Transverse sinus	⑳ Superior ophthalmic vein

Notes:-

DURAL VENOUS SINUSES

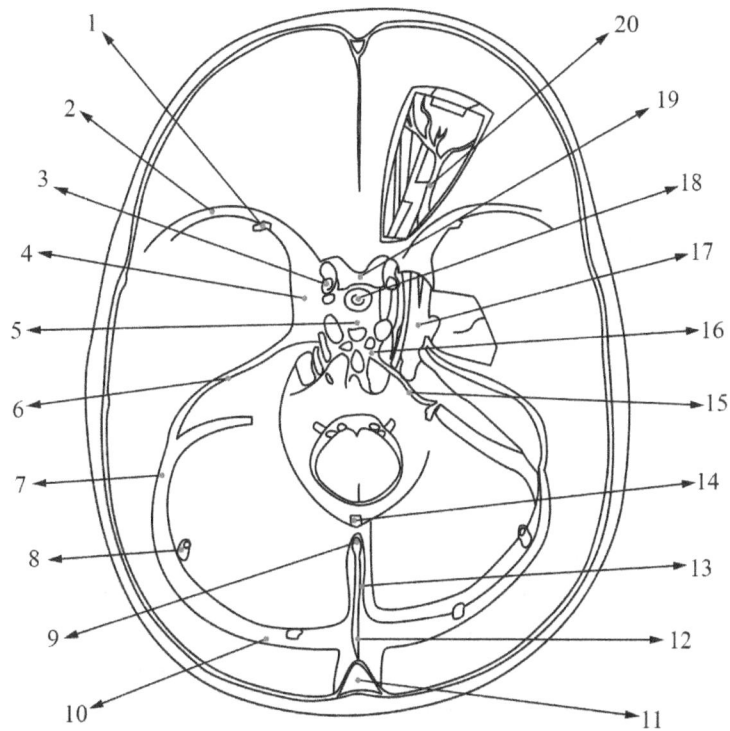

Subarachnoid cisterns of the brain

Select different colors for each of the areas of the brain provided with a color-coding circle and use them to color in the coding circles and corresponding structures in the diagram.

①	Third ventricle	⑪	Lateral aperture of fourth ventricle
②	Cerebral aqueduct	⑫	Cerebral subarachnoid space
③	Interpeduncular cistern	⑬	Quadrigeminal cistern
④	Chiasmatic cistern	⑭	Choroid plexus of third ventricle
⑤	Fourth ventricle		
⑥	Pontocerebellar cistern		
⑦	Spinal subarachnoid space		
⑧	Posterior cerebellomedullary cistern		
⑨	Median aperture of fourth ventricle		
⑩	Choroid plexus of fourth ventricle		

Notes:-

SUBARACHNOID CISTERNS OF THE BRAIN

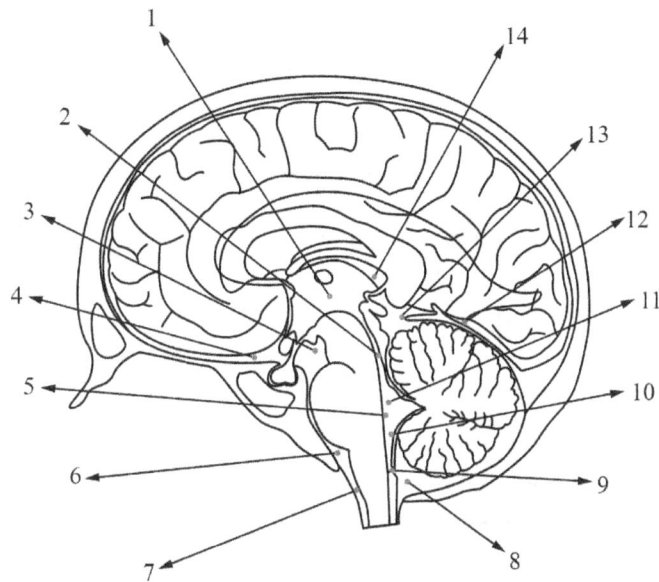

Spinal cord in SITU

Select different colors for each of the areas of the brain provided with a color-coding circle and use them to color in the coding circles and corresponding structures in the diagram.

① Body of the vertebra	⑪ Medial muscular ramus	㉑ Lung
② Subarachnoid space	⑫ Epidural space	㉒ Thoracic aorta
③ Lateral horn of spinal cord	⑬ Spinal cord	
④ White ramus communicans	⑭ Dura mater of spinal cord	
⑤ Anterior root of spinal nerve	⑮ Dorsal root ganglion	
⑥ Recurrent meningeal branches of spinal nerve	⑯ Anterior ramus of spinal nerve	
⑦ Posterior ramus of spinal nerve	⑰ Pleura	
⑧ Posterior root of spinal nerve	⑱ Gray ramus communicans	
⑨ Lateral muscular ramus	⑲ Pia mater of spinal cord	
⑩ Arachnoid mater of spinal cord	⑳ Ganglion of sympathetic trunk Pia mater of	

Notes:-

SPINAL CORD IN SITU

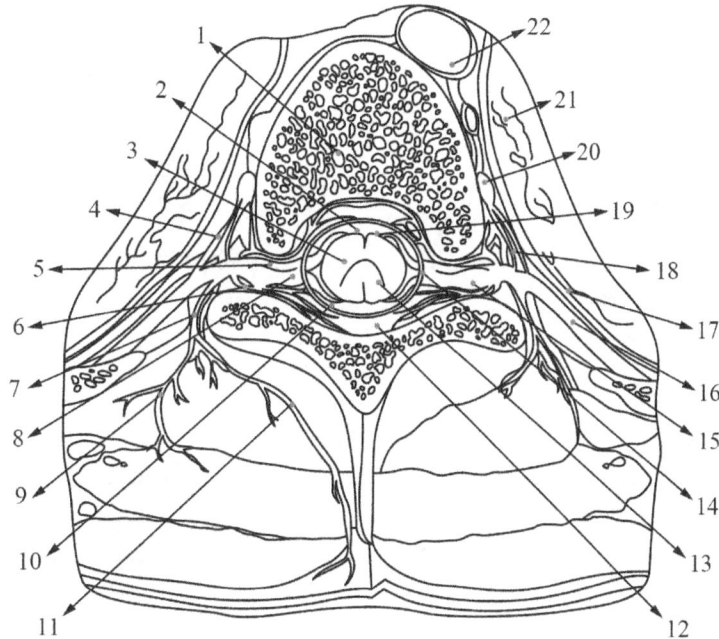

Structure of the spinal cord

Select different colors for each of the areas of the brain provided with a color-coding circle and use them to color in the coding circles and corresponding structures in the diagram.

①	Posterior median sulcus of spinal cord	⑪	Sacral nerves S1-S5
②	Ascending cervical artery	⑫	Dura mater of spinal cord
③	1st rib	⑬	Medullary cone
④	Spinal ganglion	⑭	Posterior spinal arteries
⑤	Posterior radicular arteries	⑮	Spinal cord
⑥	Spinal nerves T1-T12	⑯	Posterior roots
⑦	12th Rib	⑰	Vertebral artery
⑧	Spinal nerves L1-L4	⑱	Spinal nerves C1-C8
⑨	Cauda equina		
⑩	Coccygeal nerve		

Notes:-

STRUCTURE OF THE SPINAL CORD

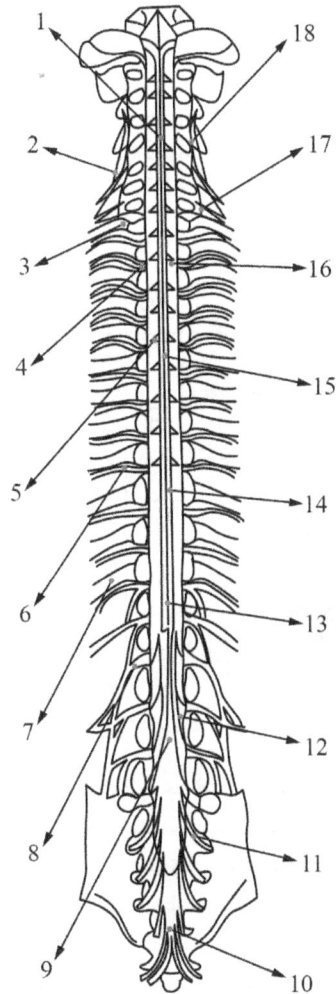

Spinal cord:Cross Section

Select different colors for each of the areas of the brain provided with a color-coding circle and use them to color in the coding circles and corresponding structures in the diagram.

①	Marginal nucleus	⑪	Lamina II	㉑	Anterior root of spinal nerve	㉛	Septomarginal fasciculus
②	Central gelatinous substance	⑫	Lamina III	㉒	Rootlets	㉜	Lateral corticospinal tract
③	Nucleus proprius	⑬	Lamina IV	㉓	Lateral funiculus	㉝	Rubrospinal tract
④	Secondary visceral grey matter	⑭	Lamina V	㉔	Gray matter	㉞	Medullary reticulospinal tract
⑤	Posterior thoracic nucleus	⑮	Lamina VI	㉕	Anterior horn	㉟	Anterior proper fasciculus
⑥	Intermediolateral nucleus	⑯	Lamina IX	㉖	Anterior median fissure of medulla	㊱	Vestibulospinal tract
⑦	Lateral motor nucleus	⑰	Lamina VII	㉗	Anterior funiculus	㊲	Anterior pontoreticulospinal tract
⑧	Lamina X	⑱	Lamina VIII	㉘	Gracile fasciculus	㊳	Tectospinal tract
⑨	Medial motor nuclei	⑲	Posterior root of spinal nerve	㉙	Interfascicular fasciculus	㊴	Anterior corticospinal tract
⑩	Lamina I	⑳	Central canal	㉚	Dorsolateral tract	㊵	Medial longitudinal fasciculus
㊶	Spinothalamic tract & spinoreticular tract	㊷	Spino-olivary tract	㊸	Anterior spinocerebellar tract	㊹	Dorsal spinocerebellar tract
㊺	Cuneate fasciculus						

Notes:-

SPINAL CORD: CROSS SECTION

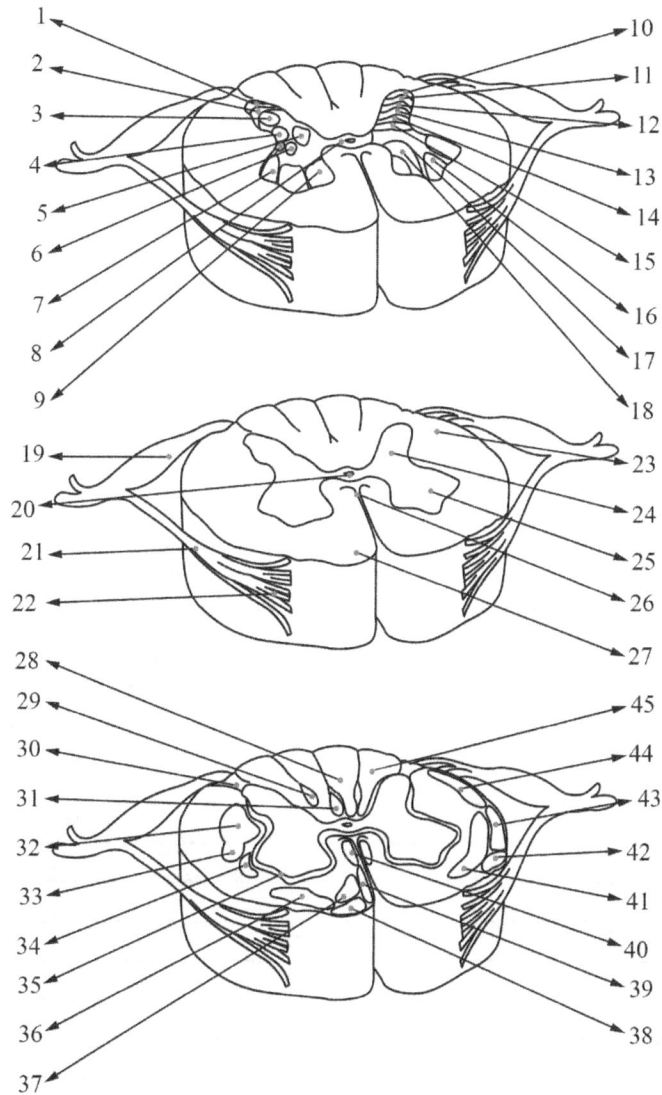

1
2
3
4
5
6
7
8
9

10
11
12
13
14
15
16
17
18

19
20
21
22

23
24
25
26
27

28
29
30
31
32
33
34
35
36
37

45
44
43
42
41
40
39
38

Spinal membranes and Nerve Roots

Select different colors for each of the areas of the brain provided with a color-coding circle and use them to color in the coding circles and corresponding structures in the diagram.

①	White matter	⑪	Dura mater of spinal cord
②	Gray matter	⑫	Anterior ramus of spinal nerve
③	Posterior root of spinal nerve	⑬	Vascular plexus of pia mater
④	Anterior spinal artery	⑭	Spinal ganglion
⑤	Spinal nerve	⑮	Gray ramus communicans
⑥	Rootlets of the ventral root	⑯	Rootlets of posterior root
⑦	Posterior ramus of spinal nerve	⑰	Anterior horn of spinal cord
⑧	Anterior root of spinal nerve	⑱	Lateral horn of spinal cord
⑨	White ramus communicans	⑲	Arachnoid mater of spinal cord
⑩	Arachnoid mater of spinal cord	⑳	Posterior horn of spinal cord

Notes:-

SPINAL MEMBRANES AND NERVE ROOTS

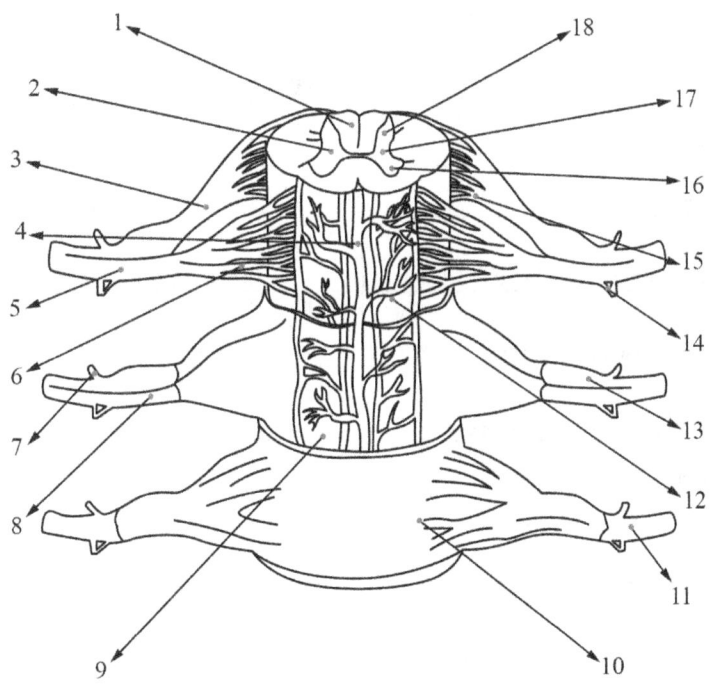

Blood Vessels of the Spinal cord

Select different colors for each of the areas of the brain provided with a color-coding circle and use them to color in the coding circles and corresponding structures in the diagram.

①	Posterior spinal vein	⑪	Dorsal branch of the posterior intercostal artery
②	Venous plexus	⑫	Thoracic aorta
③	Sulcal vein	⑬	Right posterior spinal artery
④	Anterior spinal vein	⑭	eft posterior spinal artery
⑤	Posterior radicular vein	⑮	Sulcal arteries
⑥	Anterior radicular vein	⑯	Posterior radicular artery
⑦	Lateral cutaneous branch	⑰	Anterior segmental medullary artery
⑧	Medial cutaneous branch	⑱	Arterial plexus
⑨	Posterior intercostal arteries	⑲	Anterior spinal artery
⑩	Spinal branch	⑳	Anterior radicular artery

Notes:-

BLOOD VESSELS OF THE SPINAL CORD

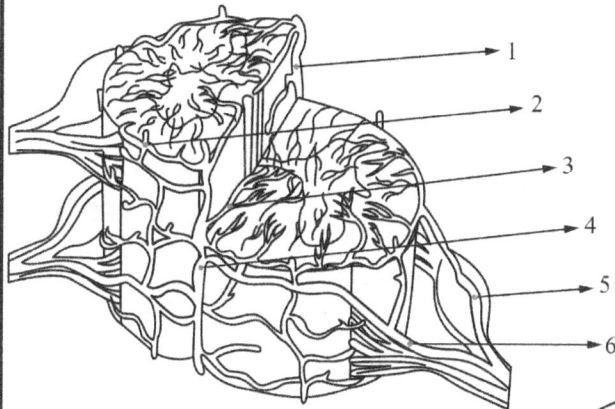

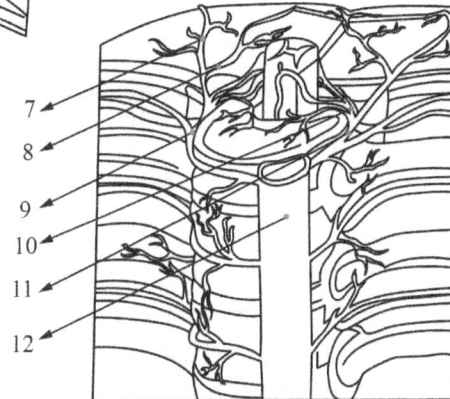

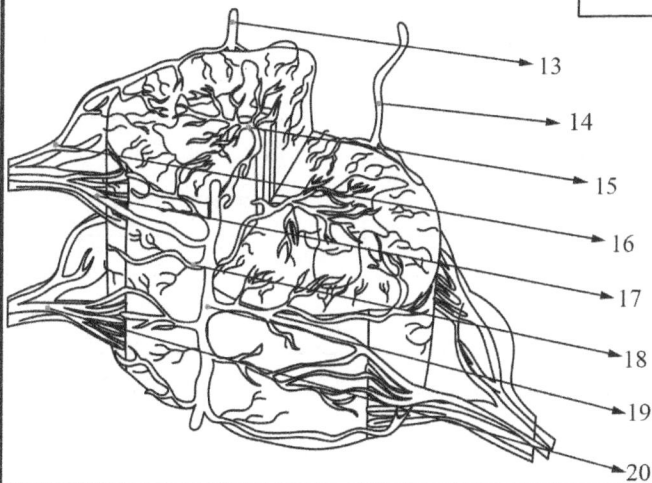

1

2

3

4

5

6

7

8

9

10

11

12

13

14

15

16

17

18

19

20

Pyramidal Tracts

Select different colors for each of the areas of the brain provided with a color-coding circle and use them to color in the coding circles and corresponding structures in the diagram.

①	Thalamus	⑪	Caudal medulla	㉑	Substantia nigra
②	Motor cortex	⑫	Spinal nerve	㉒	Putamen
③	Globus pallidus	⑬	Anterior root of spinal nerve	㉓	Caudate nucleus
④	Internal capsule	⑭	Spinal cord	㉔	Cerebral aqueduct
⑤	Claustrum	⑮	Posterior root of spinal nerve		
⑥	Corticobulbar tract	⑯	Anterior corticospinal tract		
⑦	Periaqueductal gray	⑰	Lateral corticospinal tract		
⑧	Cerebral peduncle	⑱	Pyramidal decussation		
⑨	Pons	⑲	Rostral medulla		
⑩	Olivary nuclei	⑳	Midbrain		

Notes:-

PYRAMIDAL TRACTS

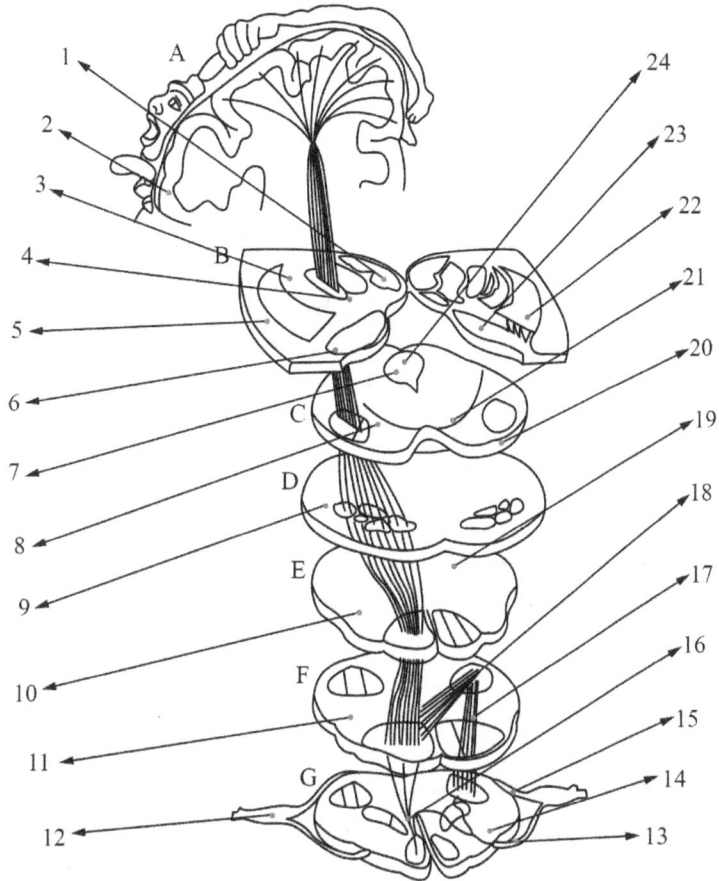

Posterior Column-Medial Leminscus Pathway(PCML)

Select different colors for each of the areas of the brain provided with a color-coding circle and use them to color in the coding circles and corresponding structures in the diagram.

①	Postcentral gyrus	⑪	Spinocervical tract
②	Ventral posterolateral nucleus of thalamus	⑫	Dorsal root ganglion
③	Periaqueductal gray	⑬	Cuneate fasciculus
④	Cerebral peduncle	⑭	Proprioception and position fibers to the cervical spinal cord
⑤	Gracile nucleus	⑮	Touch, Pressure, and vibration fibers to the cervical spinal cord
⑥	Rostral medulla	⑯	Cuneate nucleus
⑦	Gracile fasciculus	⑰	Midbrain
⑧	Lateral cervical nucleus	⑱	Substantia nigra
⑨	Cervical part of spinal cord		
⑩	Lumbar part of spinal cord		

Notes:-

POSTERIOR COLUMN–MEDIAL LEMNISCUS PATHWAY (PCML)

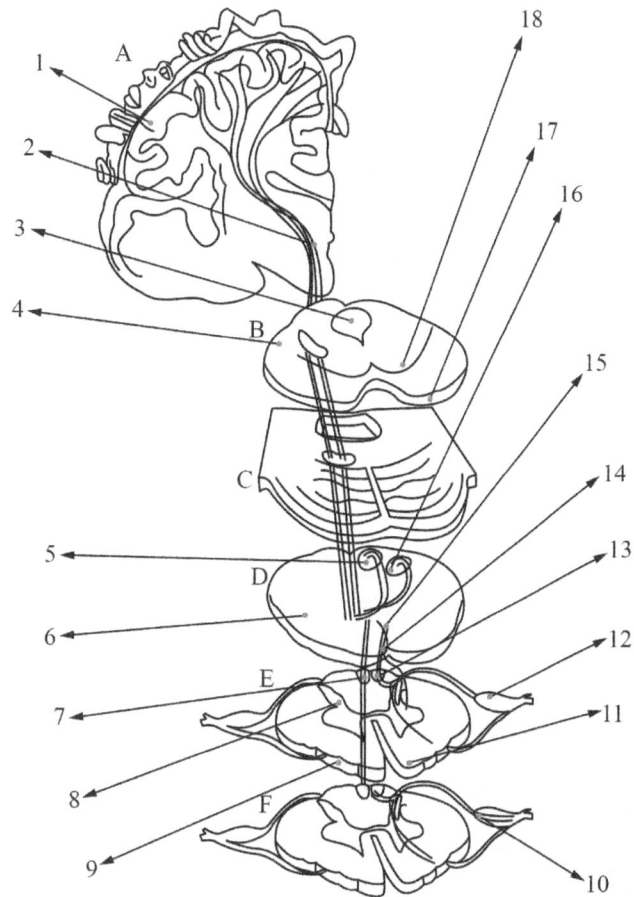

Taste Pathway

Select different colors for each of the areas of the brain provided with a color-coding circle and use them to color in the coding circles and corresponding structures in the diagram.

①	Amygdaloid body	⑪	Rostral solitary nucleus	㉑	Lingual nerve			
②	Ventral posteromedial nucleus	⑫	Petrosal ganglion	㉒	Mandibular nerve			
③	Trigeminal trunk	⑬	Nodose ganglion	㉓	Pterygopalatine ganglion			
④	Lateral hypothalamic area	⑭	Nodose ganglion	㉔	Otic ganglion			
⑤	Pontine taste area	⑮	Superior laryngeal nerve	㉕	Maxillary nerve			
⑥	Geniculate ganglion	⑯	Larynx	㉖	Greater petrosal nerve			
⑦	Facial nerve	⑰	Epiglottis	㉗	Trigeminal ganglion			
⑧	Pons	⑱	Vallate papillae	㉘	Taste sensory cortex			
⑨	Intermediate nerve	⑲	Chorda tympani					
⑩	Glossopharyngeal nerve	⑳	Foliate papillae					

Notes:-

TASTE PATHWAY

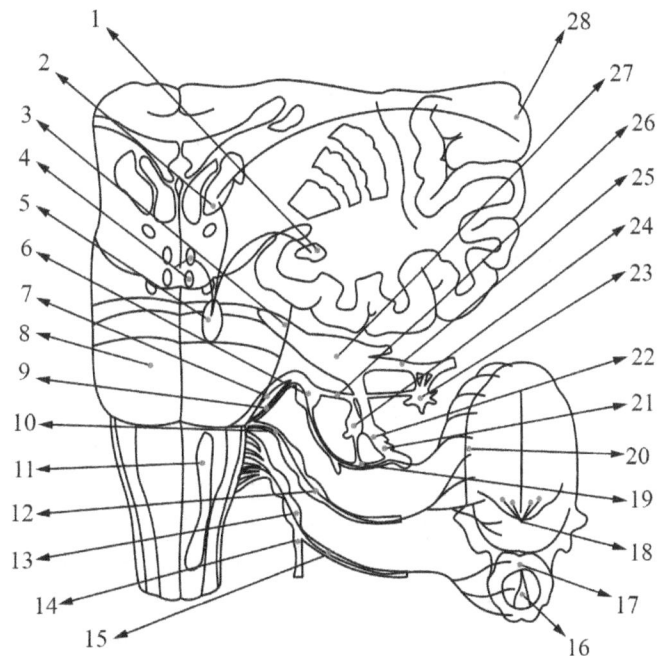

12 Cranial Nerves

Select different colors for each of the areas of the brain provided with a color-coding circle and use them to color in the coding circles and corresponding structures in the diagram.

①	Olfactory nerve	⑪	Trigeminal nerve
②	Oculomotor nerve	⑫	Optic nerve
③	Trochlear nerve		
④	Facial nerve		
⑤	Glossopharyngeal nerve		
⑥	Vagus nerve		
⑦	Accessory nerve		
⑧	Hypoglossal nerve		
⑨	Vestibulocochlear nerve		
⑩	Abducens nerve		

Notes:-

----

12 CRANIAL NERVES

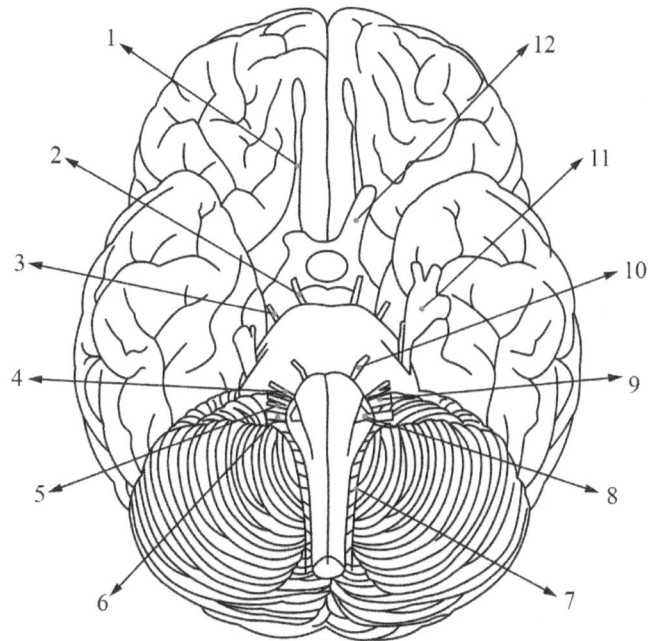

Olfactory Nerve

Select different colors for each of the areas of the brain provided with a color-coding circle and use them to color in the coding circles and corresponding structures in the diagram.

①	Paraterminal gyrus	⑪	Medial olfactory stria	㉑	Olfactory mucus layer
②	Subcallosal area	⑫	Anterior perforated substance	㉒	Olfactory cilia
③	Olfactory bulb	⑬	Uncus	㉓	Basal cells
④	Dura mater	⑭	Amygdaloid body	㉔	Lamina propria
⑤	Frontal sinus	⑮	Parahippocampal gyrus		
⑥	Cribriform plate	⑯	Efferent fibers to olfactory bulb		
⑦	Olfactory tract	⑰	Afferent fibers from olfactory bulb		
⑧	Olfactory trigone	⑱	Olfactory glomerulus		
⑨	Ambient gyrus	⑲	Olfactory glands (Bowman)		
⑩	Anterior commissure	⑳	Epithelium olfactorium		

Notes:-

OLFACTORY NERVE

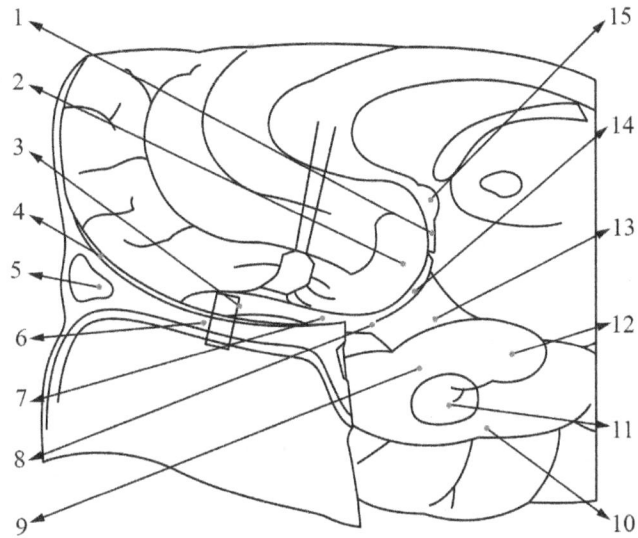

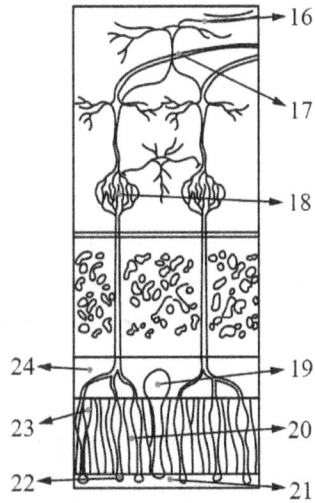

Optic Nerve

Select different colors for each of the areas of the brain provided with a color-coding circle and use them to color in the coding circles and corresponding structures in the diagram.

①	Ocular bulbs
②	Optic tract
③	Medial geniculate nucleus
④	Inferior colliculus
⑤	Optic radiation
⑥	Superior colliculus
⑦	Lateral geniculate nucleus
⑧	Optic chiasm
⑨	Optic nerve

Notes:-

OPTIC NERVE

Oculomotor, trochlear and abducens nerves

Select different colors for each of the areas of the brain provided with a color-coding circle and use them to color in the coding circles and corresponding structures in the diagram.

①	Oculomotor nerve	⑪	Oculomotor nucleus
②	Superior branch of oculomotor nerve	⑫	Accessory oculomotor nucleus
③	Short ciliary nerves		
④	Ciliary ganglion		
⑤	Parasympathetic root of ciliary ganglion		
⑥	Inferior branch of oculomotor nerve		
⑦	Abducens nerve		
⑧	Trochlear nerve		
⑨	Abducens nucleus		
⑩	Trochlear nucleus		

Notes:-

OCULOMOTOR, TROCHLEAR AND ABDUCENS NERVES

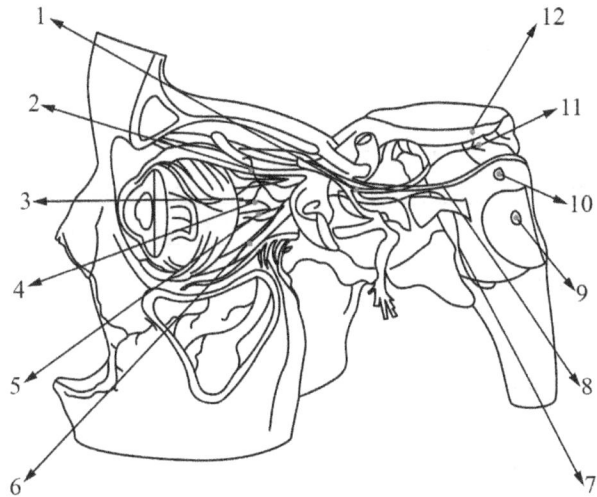

Ophthalmic nerve

Select different colors for each of the areas of the brain provided with a color-coding circle and use them to color in the coding circles and corresponding structures in the diagram.

①	Frontal nerve	⑪	Optic nerve
②	Anterior ethmoidal nerve	⑫	Ophthalmic nerve
③	Supraorbital nerve	⑬	Recurrent tentorial branch of ophthalmic nerve
④	Supratrochlear nerve	⑭	Trigeminal ganglion
⑤	Infratrochlear nerve	⑮	Trigeminal nerve
⑥	Posterior ethmoidal nerve	⑯	Internal carotid artery
⑦	Long ciliary nerves	⑰	Nasociliary nerve
⑧	Short ciliary nerves	⑱	ensory root of ciliary ganglion
⑨	Ciliary ganglion	⑲	Lacrimal nerve
⑩	Pterygopalatine ganglion		

Notes:-

OPHTHALMIC NERVE

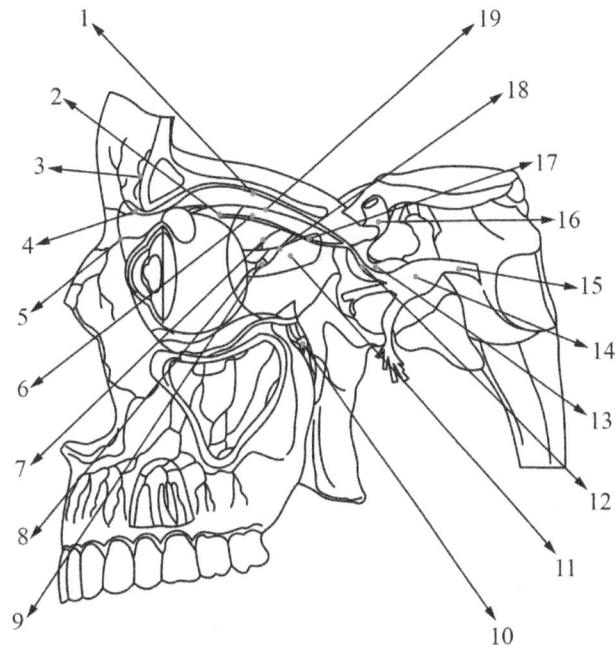

Maxillary nerve

Select different colors for each of the areas of the brain provided with a color-coding circle and use them to color in the coding circles and corresponding structures in the diagram.

①	Communicating branch of lacrimal nerve with zygomatic nerve	⑪	Pharyngeal nerve
②	Sensory root of pterygopalatine ganglion	⑫	Nasopalatine nerve
③	Zygomatic nerve	⑬	Pterygopalatine ganglion
④	Infraorbital nerve	⑭	Mandibular nerve
⑤	Posterior superior alveolar nerve	⑮	Meningeal branch of maxillary nerve
⑥	Middle superior alveolar nerve	⑯	Trigeminal nerve
⑦	Anterior superior alveolar nerve	⑰	Trigeminal ganglion
⑧	Superior dental branches	⑱	Maxillary nerve
⑨	Superior dental plexus		
⑩	Palatine nerves		

Notes:-

MAXILLARY NERVE

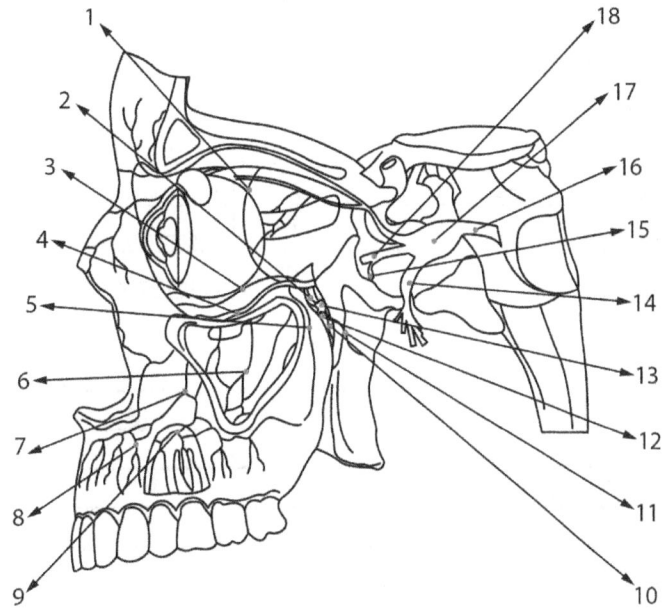

Mandibular nerve

Select different colors for each of the areas of the brain provided with a color-coding circle and use them to color in the coding circles and corresponding structures in the diagram.

①	Anterior deep temporal nerve	⑪	Inferior alveolar nerve
②	Posterior deep temporal nerve	⑫	Masseteric nerve
③	Lingual nerve	⑬	Middle meningeal artery
④	Buccal nerve	⑭	Auriculotemporal nerve
⑤	Submandibular ganglion	⑮	Meningeal branch of mandibular nerve
⑥	Sublingual nerve	⑯	Mandibular nerve
⑦	Mental foramen	⑰	Maxillary nerve
⑧	Inferior dental branches	⑱	Trigeminal nerve
⑨	Mental nerve	⑲	Trigeminal ganglion
⑩	Mylohyoid nerve	⑳	Ophthalmic nerve

Notes:-

MANDIBULAR NERVE

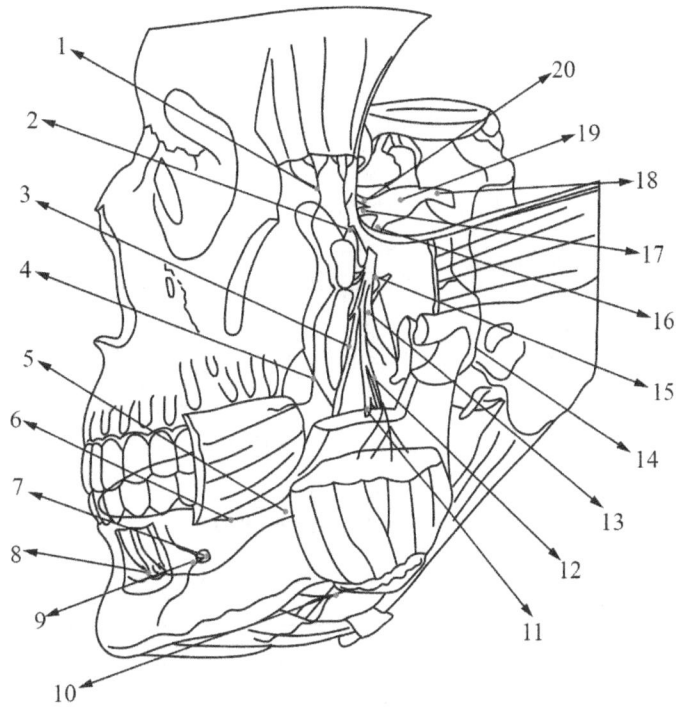

Facial nerve

Select different colors for each of the areas of the brain provided with a color-coding circle and use them to color in the coding circles and corresponding structures in the diagram.

①	Trigeminal ganglion	⑪	Submandibular ganglion	㉑	Auricular branch of posterior auricular nerve
②	Internal carotid plexus	⑫	Marginal mandibular branch of facial nerve	㉒	Tympanic plexus
③	Maxillary nerve	⑬	Cervical branch of facial nerve	㉓	Intermediate nerve
④	Greater petrosal nerve	⑭	Stylohyoid branch of the facial nerve	㉔	Lesser petrosal nerve
⑤	Pterygopalatine ganglion	⑮	Digastric branch of the facial nerve	㉕	Geniculate ganglion
⑥	Temporal branches of facial nerve	⑯	Chorda tympani	㉖	Trigeminal nerve
⑦	Zygomatic branches of facial nerve	⑰	Posterior auricular nerve		
⑧	Otic ganglion	⑱	Facial nerve		
⑨	Buccal branches of facial nerve	⑲	Glossopharyngeal nerve		
⑩	Mandibular nerve	⑳	Occipital branch of posterior auricular nerve		

Notes:-

FACIAL NERVE

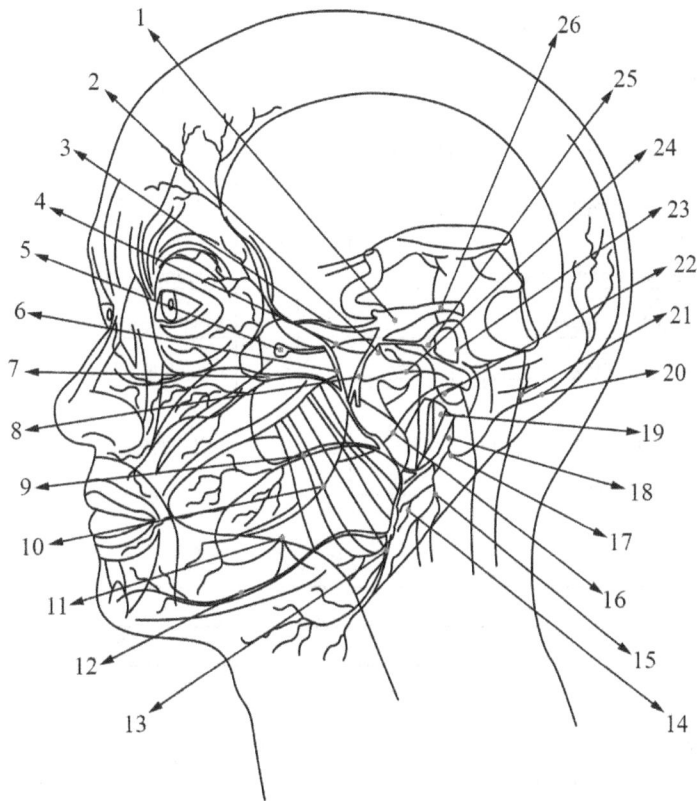

Vestibulocochlear nerve

Select different colors for each of the areas of the brain provided with a color-coding circle and use them to color in the coding circles and corresponding structures in the diagram.

①	Vestibular nerve	⑪	Posterior cochlear nucleus
②	Anterior ampullary nerve	⑫	Anterior cochlear nucleus
③	Lateral ampullary nerve	⑬	Inferior vestibular nucleus
④	Superior part of vestibular ganglion	⑭	Medial vestibular nucleus
⑤	Utricular nerve	⑮	Lateral vestibular nucleus
⑥	Posterior ampullary nerve	⑯	Superior vestibular nucleus
⑦	Chorda tympani	⑰	Vestibulocochlear nerve
⑧	Saccular nerve	⑱	Facial nerve
⑨	Inferior part of vestibular ganglion	⑲	Cochlear nerve
⑩	Spiral ganglion		

Notes:-

VESTIBULOCOCHLEAR NERVE

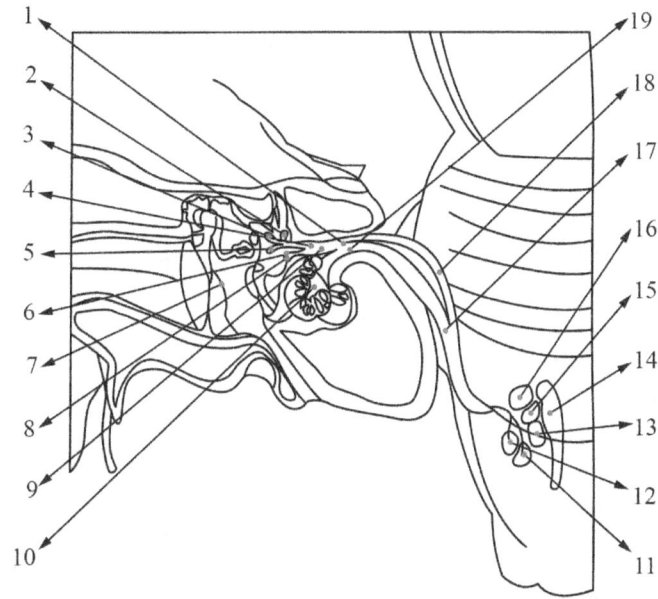

1
2
3
4
5
6
7
8
9
10

19
18
17
16
15
14
13
12
11

Glossopharyngeal nerve

Select different colors for each of the areas of the brain provided with a color-coding circle and use them to color in the coding circles and corresponding structures in the diagram.

①	Spinal nucleus and tract of trigeminal nerve	⑪	Palatine tonsil	㉑	Pharyngeal plexus of glossopharyngeal nerve
②	Pons	⑫	Tongue	㉒	Lingual branches of glossopharyngeal nerve
③	Superior salivatory nucleus	⑬	Glossopharyngeal nerve	㉓	Medulla oblongata
④	Superior ganglion of glossopharyngeal nerve	⑭	Pharyngeal branches of glossopharyngeal nerve	㉔	Communicating branch of glossopharyngeal nerve with auricular branch of vagus nerve
⑤	Parotid gland	⑮	Tonsillar branches of glossopharyngeal nerve	㉕	Jugular foramen
⑥	Stylomastoid foramen	⑯	Carotid sinus	㉖	Nucleus ambiguu
⑦	Auditory tube	⑰	Carotid body	㉗	Solitary nucleus and tract
⑧	Carotid canal	⑱	Internal carotid artery	㉘	Inferior salivatory nucleus
⑨	Inferior ganglion of glossopharyngeal nerve	⑲	Pharyngeal constrictor muscle		
⑩	Communicating branch of facial nerve with glossopharyngeal nerve	⑳	Carotid branch of glossopharyngeal nerve		

Notes:-

GLOSSOPHARYNGEAL NERVE

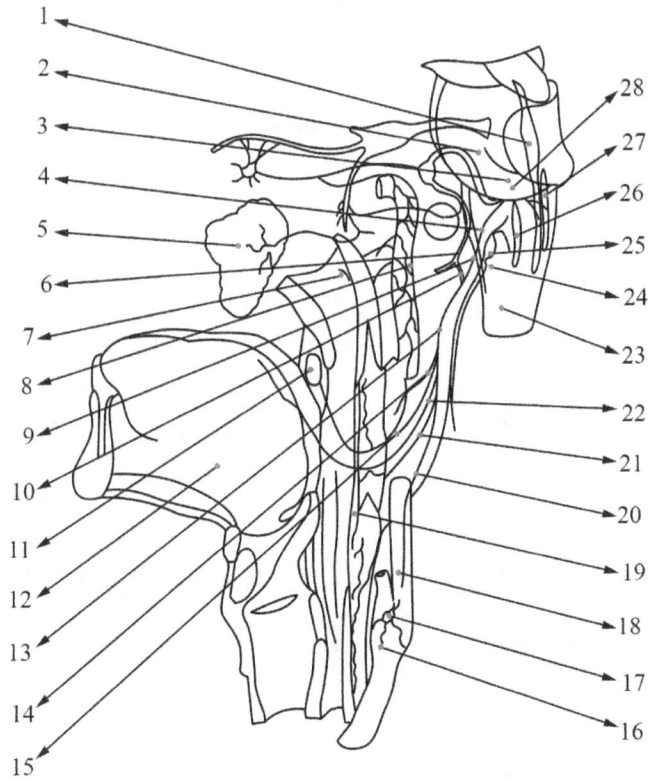

1
2
3
4
5
6
7
8
9
10
11
12
13
14
15

28
27
26
25
24
23
22
21
20
19
18
17
16

Vagus nerve

Select different colors for each of the areas of the brain provided with a color-coding circle and use them to color in the coding circles and corresponding structures in the diagram.

① Glossopharyngeal nerve	⑪ Left recurrent laryngeal nerve	㉑ Celiac branches of posterior vagal trunk
② Vagus nerve	⑫ Cardiac plexus	㉒ Esophageal plexus
③ Jugular foramen	⑬ Hepatic branch of anterior vagal trunk	㉓ Anterior vagal trunk
④ Accessory nerve	⑭ Hepatic plexus	㉔ **Inferior cardiac nerve**
⑤ Pharyngeal branches of glossopharyngeal nerve	⑮ Pancreatic plexus	㉕ Superior cervical cardiac branch of vagus nerve
⑥ Pharyngeal branch of vagus nerve	⑯ Intestinal branch of vagus nerve	㉖ Auricular branch of vagus nerve
⑦ Superior laryngeal nerve	⑰ Intermesenteric plexus	㉗ Spinal trigeminal nucleus
⑧ Internal branch of superior laryngeal nerve	⑱ Superior mesenteric ganglion	㉘ Nucleus ambiguus
⑨ External branch of superior laryngeal nerve	⑲ Splenic plexus	㉙ Dorsal nucleus of vagus nerve
⑩ Right recurrent laryngeal nerve	⑳ Anterior gastric branches of anterior vagal trunk	㉚ Nucleus of solitary tract

Notes:-

VAGUS NERVE

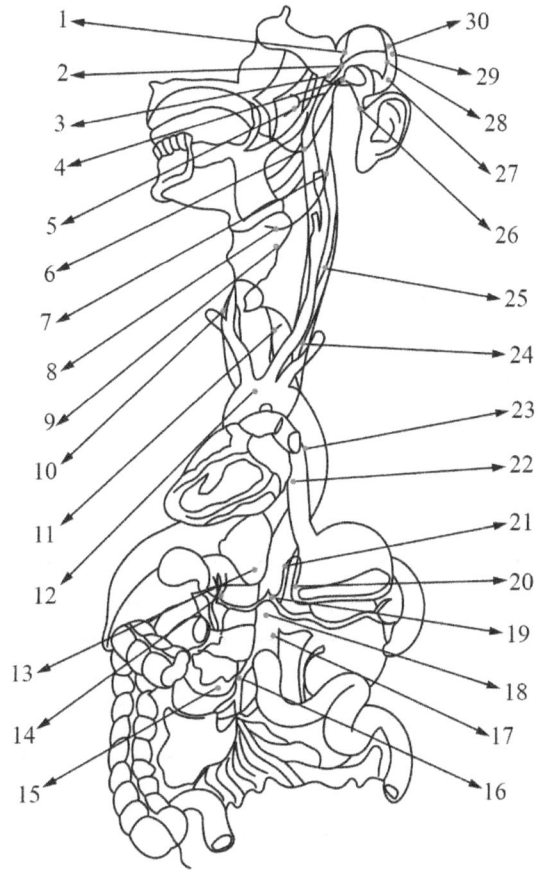

1
2
3
4
5
6
7
8
9
10
11
12
13
14
15

30
29
28
27
26
25
24
23
22
21
20
19
18
17
16

Accessory nerve

Select different colors for each of the areas of the brain provided with a color-coding circle and use them to color in the coding circles and corresponding structures in the diagram.

①	Vagus nerve	⑪	Sternocleidomastoid muscle
②	Cranial root of accessory nerve	⑫	Spinal nerve Cl
③	Nucleus ambiguus	⑬	Inferior ganglion of vagus nerve
④	Spinal root of accessory nerve	⑭	Jugular foramen
⑤	Spinal nerve C2	⑮	Superior ganglion of vagus nerve
⑥	Spinal nerve C3		
⑦	Spinal nerve C4		
⑧	**Spinal nerve C5**		
⑨	Trapezius muscle		
⑩	Accessory nerve		

Notes:-

ACCESSORY NERVE

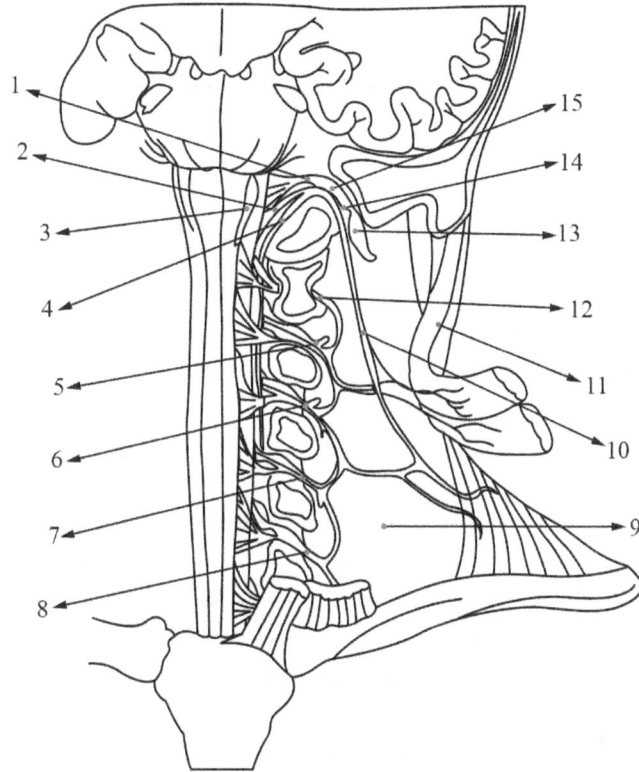

Hypoglossal nerve

Select different colors for each of the areas of the brain provided with a color-coding circle and use them to color in the coding circles and corresponding structures in the diagram.

①	Tongue	⑪	Superior root of ansa cervicalis
②	Styloglossus muscle	⑫	Ventral roots of spinal nerves C1-C3
③	Genioglossus muscle	⑬	Hypoglossal nerve
④	Hyoglossus muscle	⑭	Hypoglossal nucleus
⑤	Thyrohyoid muscle		
⑥	Sternohyoid muscle		
⑦	Omohyoid muscle		
⑧	Sternothyroid muscle		
⑨	Ansa cervicalis		
⑩	Inferior root of ansa cervicalis		

Notes:-

HYPOGLOSSAL NERVE

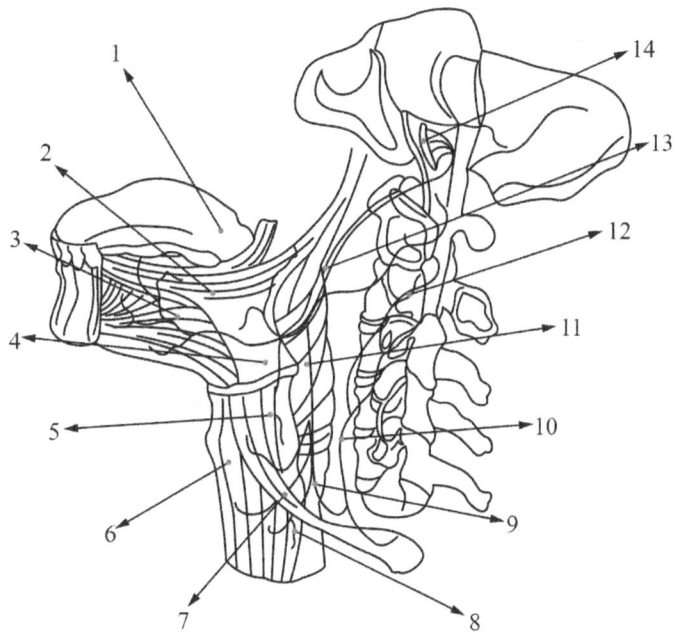

Vertebral column and spinal nerves

Select different colors for each of the areas of the brain provided with a color-coding circle and use them to color in the coding circles and corresponding structures in the diagram.

①	Spinal nerves C1 C8
②	Spinal nerves T1-T12
③	Spinal nerves L1-L5
④	Spinal nerves S1 S5
⑤	Cervical enlargement
⑥	Lumbar enlargement
⑦	Medullary cone
⑧	Coccygeal nerve
⑨	Filum terminale

Notes:-

VERTEBRAL COLUMN AND SPINAL NERVES

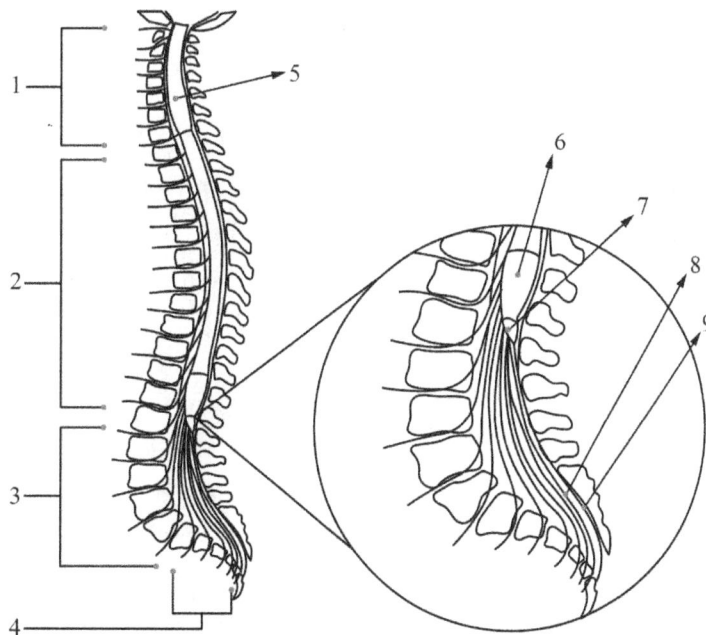

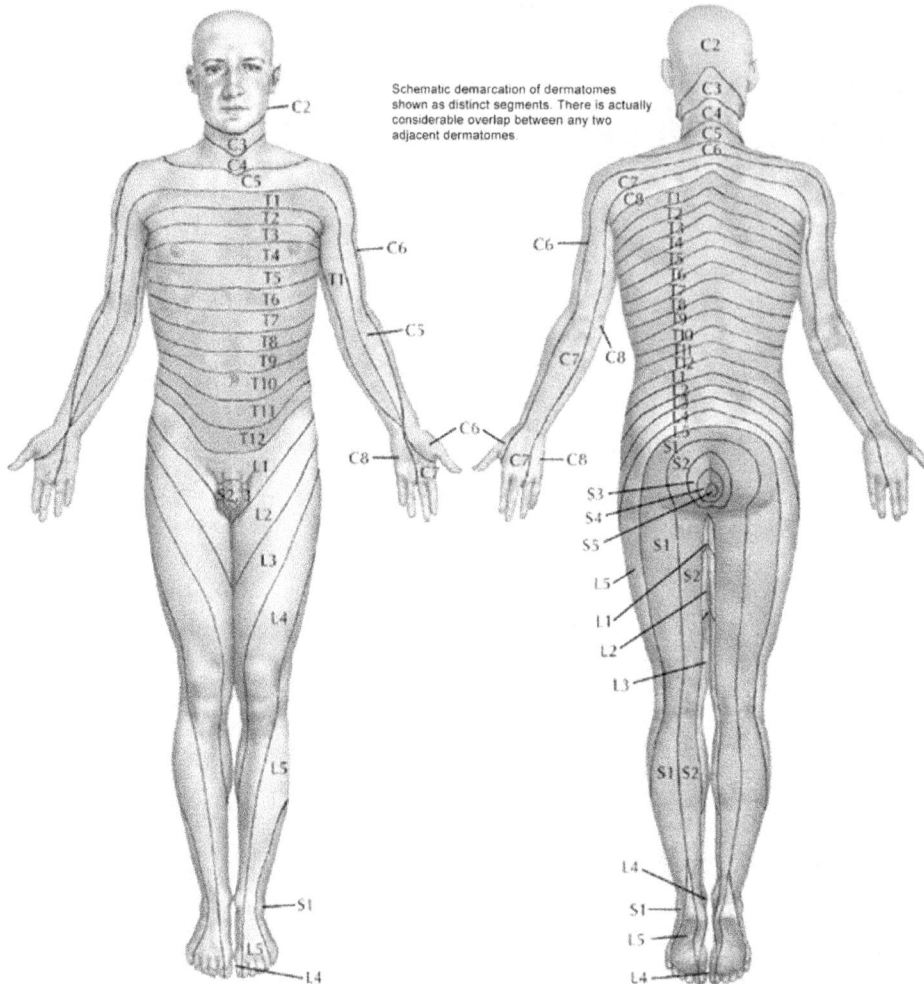

Schematic demarcation of dermatomes
shown as distinct segments. There is actually
considerable overlap between any two
adjacent dermatomes.

Levels of principal dermatomes

C5	Clavicles	T10	Level of umbilicus
C5,6,7	Lateral parts of upper limbs	T12	Inguinal or groin regions
C8, T1	Medial sides of upper limbs	L1,2,3,4	Anterior and inner surfaces of lower limbs
C6	Thumb	L4,5, S1	Foot
C6,7,8	Hand	L4	Medial side of great toe
C8	Ring and little fingers	S1, 2, L5	Posterior and outer surfaces of lower limbs
T4	Level of nipples	S1	Lateral margin of foot and little toe
		S2,3,4	Perineum

DERMATOMES

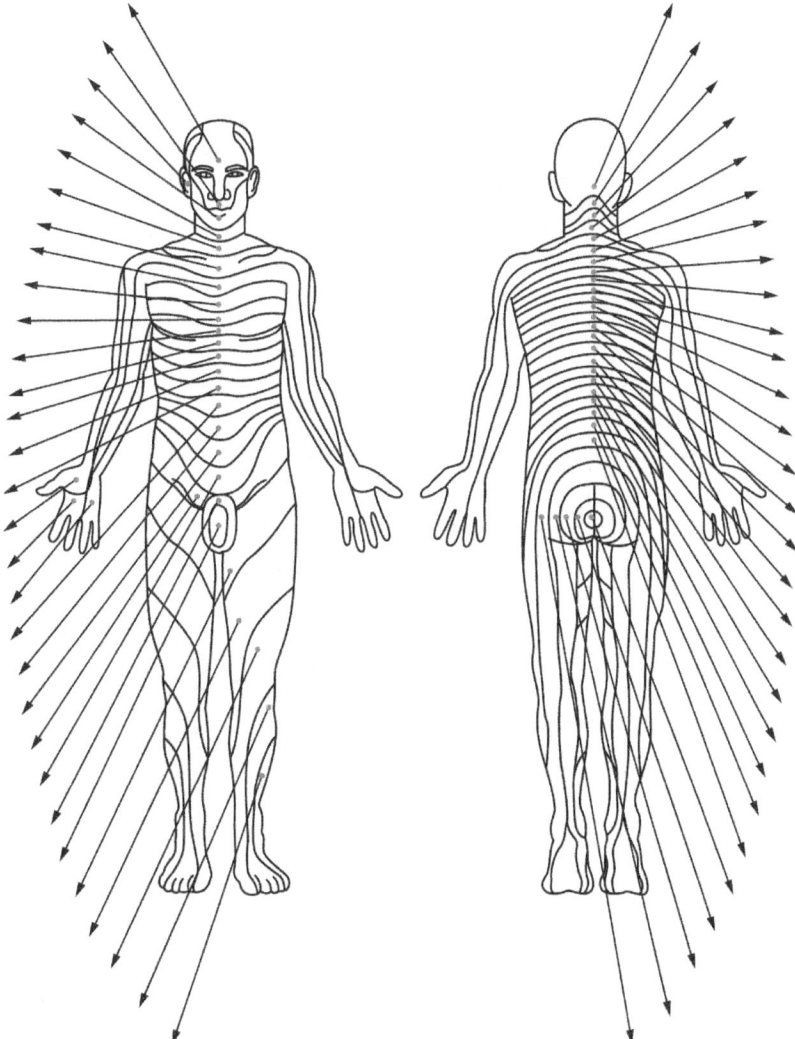

Autonomic nervous system

Select different colors for each of the areas of the brain provided with a color-coding circle and use them to color in the coding circles and corresponding structures in the diagram.

①	Superior Cervical ganglion	⑪	Inferior mesenteric ganglion	㉑	Facial nerve
②	Cervical part of sympathetic trunk	⑫	Lumbar splanchnic nerves	㉒	Oculomotor nerve
③	Middle cervical ganglion	⑬	Sympathetic trunk		
④	Inferior cervical ganglion	⑭	Pelvic splanchnic nerves		
⑤	Greater thoracic splanchnic nerve	⑮	Otic ganglion		
⑥	Celiac ganglia	⑯	Submandibular ganglion		
⑦	Aorticorenal ganglia	⑰	Pterygopalatine ganglion		
⑧	Lesser thoracic splanchnic nerve	⑱	Ciliary ganglion		
⑨	Least thoracic splanchnic nerve	⑲	Vagus nerve		
⑩	Superior mesenteric ganglion	⑳	Glossopharyngeal nerve		

Notes:-

AUTONOMIC NERVOUS SYSTEM

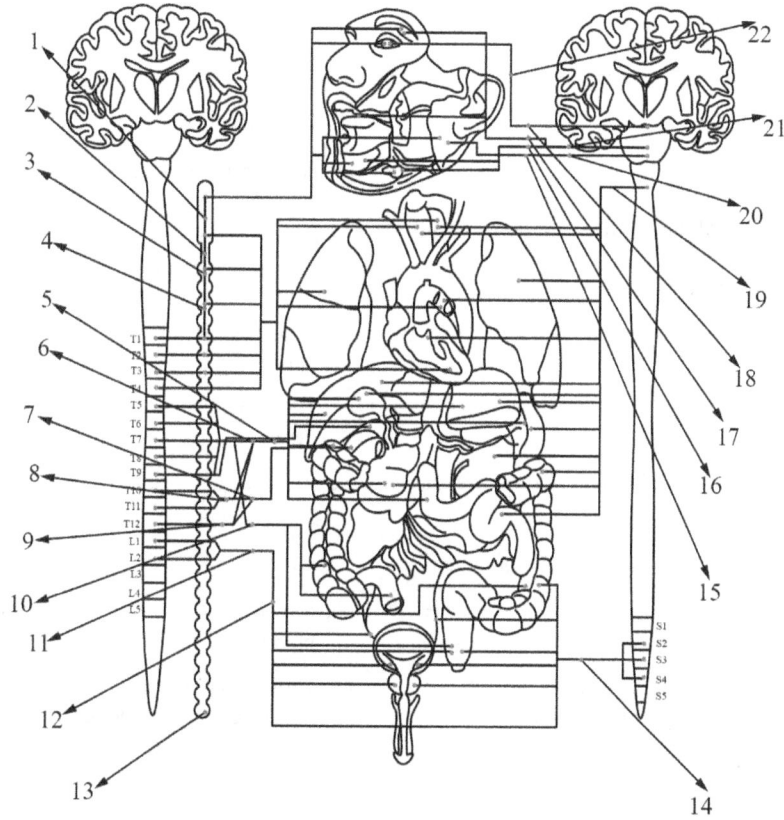

1

2

3

4

5

6

7

8

9

10

11

12

13

14

15

16

17

18

19

20

21

22

T1
T2
T3
T4
T5
T6
T7
T8
T9
T10
T11
T12
L1
L2
L3
L4
L5

S1
S2
S3
S4
S5

1. Which part of the brain has a blood-brain barrier?

   (A) posterior pituitary
   (B) median eminence of the hypothalamus
   (C) anterior pituitary
   (D) pineal body
   (E) area postrema of the fourth ventricle

2. Cell bodies for the motor supply of the trigeminal nerve lie

   (A) Midbrain
   (B) Floor of fourth ventricle
   (C) Hypothalamus
   (D) Cerebral cortex
   (E) Posterior to cerebral aqueduct

3. Cell bodies for the motor supply of the facial nerve lie

   (A) Midbrain
   (B) Hypothalamus
   (C) Pons
   (D) Floor of the third ventricle
   (E) None of the above

4. The lumbar plexus

   (A) The femoral nerve is formed from L2, 3, 4
   (B) Is derived from the last three lumbar nerves
   (C) Is formed from the posterior rami
   (D) The pudendal nerve is a branch of the lumbar plexus
   (E) Is immediately medial to the inferior vena cava

5. regarding the innervation of the bladder

   (A) bladder distension sensation travels with the sympathetic nervous system
   (B) sympathetic fibres are excitatory to the bladder
   (C) bladder pain travels only with the superior hypogastric plexus
   (D) parasympathetic innervation is via the pelvic splanchnic nerves
   (E) sympathetic innervation comes from L3 and L4 segments of the cord

6. With regards to the spinal cord blood supply

(A) There are two anterior spinal arteries

(B) The posterior spinal artery is singular

(C) The anterior spinal artery arises form the vertebral artery

(D) The anterior spinal artery retains a uniform size throughout its length

(E) The posterior spinal artery arises from the posterior superior cerebellar

7. The diameter of a motor nerve fibre is

(A) 3-5 micrometre

(B) 5-12 micrometre

(C) 1-2 micrometre

(D) 12-20 micrometre

(E) 20-50 micrometre

8. With regard to dermatomal nerve supply:

(A) The heel is supplied by S2

(B) The umbilicus is supplied by either T12 or L1

(C) T6 is at the level of the nipple

(D) C7 supplies the index finger

(E) The anterior axial line of the upper limb runs between C6 and C7

9. with regard to myotomal nerve supply

(A) opponens pollicis is C8

(B) shoulder abduction is C5, 6

(C) ankle plantar flexion is L4, 5

(D) elbow extension is C7, 8

(E) ankle eversion is L4

10. The afferent path of the sneeze reflex is mediated via the

(A) Maxillary nerve V2

(B) Glossopharyngeal nerve

(C) Vagus nerve

(D) Ophthalmic nerve V1

(E) Mandibular nerve V3

11. The motor nuclei of the facial nerve are situated in the

(A) Medulla oblongata

(B) Cerebellum

(C) Midbrain

(D) Floor of the third ventricle

(E) Pons

12. The dermatome supplying the great toe is usually

(A) S1

(B) L4

(C) L5

(D) L3

(E) S2

13. Regarding the cranial nerves

(A) The abducens nerve traverses the foramen lacerum

(B) The facial nerve may be involved in infection in the cavernous sinus

(C) The trigeminal nerve is purely sensory

(D) The hypoglossal nerve exits the skull through the foramen magnum

(E) The trochlear nerve supplies the superior oblique muscle only

14. Which of the following about the facial nerve is incorrect?

(A) Gives the great petrosal nerve

(B) Contains fibres destined for the ciliary ganglion

(C) Supplie buccinator

(D) Supplies muscles of facial expression

(E) Contains taste fibres

15. Which is the largest branch of the internal carotid artery?

(A) middle cerebral artery

(B) ophthalmic artery

(C) striate artery

(D) posterior communicating artery

(E) anterior cerebral artery

16. The brain stem does NOT include the:

(A) diencephalons

(B) pons

(C) subtantia nigra

(D) midbrain

(E) medilla oblongata

17. Which cranial nerve lies in the junction between pons and medilla?

(A) vestibulocochlear nerve (VIII)

(B) facial nerve (VII)

(C) abducent nerve (VI)

(D) glossopharyngeal nerve (IX)

(E) vagus nerve (X)

18. Which is the smallest cranial nerve?

(A) occulomotor nerve (III)   (B) trochlear nerve (IV)

(C) accessory nerve (XI)   (D) abducent nerve (VI)

(E) olfactory nerve (I)

19. Which midbrain cells are involved in general light reflexes?

(A) inferior colliculus   (B) medial geniculate body

(C) substantia nigra   (D) superior colliculus

(E) red nucleus

20. The medilla oblongata:

(A) lies between the midbrain and pons   (B) has only one cranial nerve emerging from it (the trigeminal nerve)

(C) passes through the foramen magnum   (D) has pyramids lateral to the olives

(E) receives its blood supply from the internal carotid artery

21. Which structure does NOT receive supply from the occulomotor nerve?

(A) levator palpebrae superioris   (B) inferior oblique

(C) ciliary body   (D) lateral rectus

(E) medial rectus

22. In central cord syndrome there is:

(A) no loss of motor or sensory function   (B) intact touch sensation with loss of all motor and other sensory functions

(C) paralysis and loss of touch sensation on one side and loss of pain and temperature sensation in the upper limbs and spasticity of the lower limbs   (D) loss of movement and all sensation below the injured segment

(E) none of the above

23. Which structure is encircled by the circle of Willis?

(A) pituitary stalk   (B) pineal gland

(C) aqueduct of the midbrain (D) cavernous sinus

(E) medulla

24. In the spinal cord:

(A) The blood supply at each level is in danger because of poor anastomoses

(B) The lateral corticospinal tract is an important motor tract

(C) Hemisection of the cord (Brown-Sequard Syndrome) results in paralysis and loss of touch and proprioception on the same side and loss of pain and temperature sensation on opposite side

(D) The cord ends at L3

(E) The dorsal/posterior columns contain primarily motor fibres

25. The vagus nerve

(A) Arises from the medulla as a single nerve

(B) Receives nucleus ambiguous fibres from the accessory nerve

(C) Supplies motor fibres to the diaphragm

(D) Supplies sensory fibres to the facial region

(E) Can be tested by looking at tongue movements

26. The sensory root of the facial nerve

(A) Emerges from the base of the skull through the foramen ovale

(B) Supplies the mucous membrane of the posterior third of the tongue

(C) Is called the nervus intermedius

(D) Presents as a swelling in the bend called the otic ganglion

(E) Arises from the sulcus between the pons and medulla

27. Which of the following is not a branch of the trigeminal nerve?

(A) Mental nerve

(B) Auricolotemporal nerve

(C) Great auricular nerve

(D) Supraorbital nerve

(E) Lacrimal nerve

28. The 4th cranial nerve supplies

(A) Medial rectus

(B) Superior oblique

(C) Orbicularis oris

(D) Inferior oblique

(E) Lateral rectus

29. The trigeminal nerve

(A) Does not carry autonomic nerves

(B) Has its motor nucleus in the upper pons

(C) Has five divisions

(D) Mandibular division is purely sensory

(E) Exits the skull entirely through the foramen ovale

30. The cervical sympathetic trunk

(A) Lies behind the prevertebral fascia

(B) Lies behind the carotid sheath

(C) Descends from the upper posterior triangle to the first rib

(D) Runs lateral to the vertebral artery

(E) Ends at the inferior cervical ganglion

31. Myotome of shoulder abduction?

(A) C5,6

(B) C5,6,7

(C) C6, 7, 8

(D) C6,7

(E) C5

32. All of the following are branches of the ophthalmic division of the trigeminal nerve EXCEPT:

(A) Supraorbital nerve

(B) Infraorbital nerve

(C) Infratrochlear nerve

(D) Lacrimal nerve

(E) Supratrochlear nerve

33. Which of the following is a branch of the mandibular nerve

(A) External nasal nerve

(B) Zygomaticofacial nerve

(C) Infraorbital nerve

(D) Zygomaticotemporal nerve

(E) Auricolotemporal nerve

34. which of the following is a branch of the maxillary nerve?

   (A) Zygomaticotemporal nerve   (B) Auricolotemporal nerve

   (C) Zygomaticofacial nerve   (D) External nasal nerve

   (E) Infraorbital nerve

35. The midbrain

   (A) Is largely in the middle cranial fossa   (B) Contains the oculomotor nuclei

   (C) Is supplied by the anterior inferior   (D) Contains the trigeminal nuclei
       cerebellar artery

   (E) Lies between pons and upper spinal
       cord

36. Cerebrospinal fluid communicates with the subarachnoid space via the

   (A) tela choroidia   (B) choroids plexus

   (C) subarachnoid granulations   (D) 3rd ventricle

   (E) 4th ventricle

37. The infratrochlear nerve supplies the

   (A) Upper lip   (B) Upper incisors

   (C) Skin of the lower eyelid   (D) Labial gum

   (E) Bridge over the nose

38. Which nerve supplies the vertex of the scalp

   (A) Supraorbital   (B) Third occipital

   (C) Auriculotemporal   (D) Greater occipital

   (E) Supratrochlear

39. Corneal sensation synapses in which ganglion

   (A) Pterygopalatine   (B) Ciliary

   (C) Otic   (D) Trigeminal

   (E) Geniculate

40. Regarding the speech centres

(A) Damage to Broca's area produces motor aphasia

(B) Broca's area is posterior

(C) Damage to Wernicke's area produces expressive aphasia

(D) Wernicke's area controls motor response

(E) Broca's area is on the left side in most left-handed people

41. Regarding the optic pathways

(A) Combined inferior rectus and superior oblique gives lateral gaze

(B) Superior rectus makes eye turn up and out

(C) Trochlear paralysis, eye cannot look downwards when turned out

(D) Abducent paralysis makes eye turn down and out

(E) Combined superior rectus and inferior oblique causes vertical upward gaze

42. Regarding the blood supply of the cerebral cortex

(A) Anterior cerebral is contralateral leg, auditory and speech

(B) Middle cerebral is contralateral arm, leg and speech areas

(C) Posterior cerebral is ipsilateral vision

(D) Middle cerebral is ipsilateral arm, face and vision

(E) Anterior cerebral is contralateral leg, micturition and defaecation

43. The septum of the nasal cavity is innervated by

(A) Nasopalatine nerve from cranial nerve V2

(B) Posterior ethmoidal nerve from V1

(C) Greater palatine nerve from V2

(D) Lesser palatine nerve from V2

(E) None of the above

44. The fifth cranial nerve supplies

(A) The conjunctiva beneath the lower eyelid via the ophthalmic nerve

(B) Skin of the earlobe via the auriculotemporal nerve

(C) Skin of the tip of the nose via the external nasal branch of the maxillary nerve

(D) Temporalis

(E) Skin over the occiput

45. The cutaneous innervation of the ear

(A) Includes the vagus

(B) Involves the greater occipital nerve

(C) Is the lesser auricular nerve

(D) Includes the zygomaticotemporal branch of the trigeminal nerve

(E) Involves the dermatome of C3

46. The ophthalmic division of the trigeminal nerve

(A) Enters the face via the inferior orbital fissure

(B) Controls abduction of the eye

(C) Supplies sensation to the forehead and upper eyelid, excluding the orbit

(D) Supplies sympathetic fibres to constrictor papillae muscles

(E) Gives five branches, two of which contain sympathetic as well as sensory fibres

47. Where does the superior cerebral vein lie?

(A) In the margins of the falx

(B) Between the dura and the skull

(C) Deep in the sulci

(D) With the superior cerebral artery

(E) In the arachnoid mater

48. Regarding the circle of Willis

(A) Internal carotid gives off ophthalmic artery

(B) Anterior communicating unites middle and anterior cerebral

(C) Posterior cerebral is a branch of the internal carotid

(D) Middle cerebral supplies motor but not sensory cortex

(E) Anterior cerebral is the largest branch of the internal carotid

49. Regarding anterior nerve roots

(A) They contain efferent fibres only

(B) All roots contain sympathetic fibres

(C) There are 31 pairs of anterior nerve roots

(D) All roots contain efferent motor fibres

(E) Anterior roots join with posterior roots 1cm distal to the intervertebral foramen

50. The following muscles are supplied by posterior rami of spinal nerves EXCEPT

(A) Splenius

(B) Semispinalis capitis

(C) Scalene posterior

(D) Levator costae

(E) Erector spinae

1. (A) (B) **(C)** (D) (E)
2. (A) **(B)** (C) (D) (E)
3. (A) (B) **(C)** (D) (E)
4. **(A)** (B) (C) (D) (F)
5. (A) (B) (C) **(D)** (E)
6. (A) (B) **(C)** (D) (E)
7. (A) (B) (C) **(D)** (E)
8. (A) (B) (C) **(D)** (E)
9. (A) (B) (C) **(D)** (E)
10. **(A)** (B) (C) (D) (E)
11. (A) (B) (C) (D) **(E)**
12. (A) (B) **(C)** (D) (E)
13. (A) (B) (C) (D) **(E)**
14. (A) **(B)** (C) (D) (E)
15. **(A)** (B) (C) (D) (E)
16. (A) (B) (C) (D) **(E)**
17. (A) (B) **(C)** (D) (E)
18. (A) **(B)** (C) (D) (E)
19. (A) (B) (C) **(D)** (E)

20. (A) (B) **(C)** (D) (E)
21. (A) (B) (C) **(D)** (E)
22. (A) **(B)** (C) (D) (E)
23. **(A)** (B) (C) (D) (E)
24. (A) **(B)** **(C)** (D) (E)
25. (A) **(B)** (C) (D) (E)
26. (A) (B) **(C)** (D) (E)
27. (A) (B) **(C)** (D) (E)
28. (A) **(B)** (C) (D) (E)
29. (A) **(B)** (C) (D) (E)
30. (A) (B) (C) **(D)** (E)
31. (A) (B) (C) (D) **(E)**
32. (A) (B) **(C)** (D) (E)
33. (A) (B) (C) (D) **(E)**
34. **(A)** (B) (C) (D) (E)
35. (A) **(B)** (C) (D) (E)
36. (A) (B) (C) (D) **(E)**
37. (A) (B) (C) (D) **(E)**
38. (A) (B) (C) **(D)** (E)

39. Ⓐ Ⓑ Ⓒ ⬤D Ⓔ

40. ⬤A Ⓑ Ⓒ Ⓓ Ⓔ

41. Ⓐ Ⓑ ⬤C Ⓓ Ⓔ

42. Ⓐ Ⓑ Ⓒ Ⓓ ⬤E

43. ⬤A Ⓑ Ⓒ Ⓓ Ⓔ

44. Ⓐ Ⓑ Ⓒ ⬤D Ⓔ

45. ⬤A Ⓑ Ⓒ Ⓓ Ⓔ

46. Ⓐ Ⓑ Ⓒ Ⓓ ⬤E

47. Ⓐ Ⓑ Ⓒ Ⓓ ⬤E

48. ⬤A Ⓑ Ⓒ Ⓓ Ⓔ

49. Ⓐ ⬤B Ⓒ Ⓓ Ⓔ

50. Ⓐ Ⓑ ⬤C Ⓓ Ⓔ

www.ingramcontent.com/pod-product-compliance
Lightning Source LLC
Chambersburg PA
CBHW081808200326
41597CB00023B/4188